William B. Dalby

Lectures on Diseases and Injuries of the Ear

Delivered at St. George's Hospital. Second Edition

William B. Dalby

Lectures on Diseases and Injuries of the Ear
Delivered at St. George's Hospital. Second Edition

ISBN/EAN: 9783337041816

Printed in Europe, USA, Canada, Australia, Japan

Cover: Foto ©berggeist007 / pixelio.de

More available books at **www.hansebooks.com**

LECTURES

on

DISEASES AND INJURIES
OF THE EAR

DELIVERED AT

ST GEORGE'S HOSPITAL

BY

W. B. DALBY, F.R.C.S., M.B. Cantab.

AURAL SURGEON TO THE HOSPITAL

SECOND EDITION

LONDON
J. & A. CHURCHILL, NEW BURLINGTON STREET
1880

PREFACE

SECOND EDITION

THE publication of these lectures met with so favorable a reception from the profession that I regret the issue of a second edition has been unavoidably delayed so long after the first has been exhausted.

With few exceptions in the present edition, the original text has been adhered to.

In the lecture upon the external ear, a brief notice has been given to branchial fistula. Some further remarks are added upon the extraction of foreign bodies from the ear, and the subjects of bony growths in the external canal has been dealt with in a more exhaustive manner. Whatever alterations or addition are found under the heading of diseases of the middle ear will give evidence of an increasing belief in the importance of the tympanic membrane as a protective membrane, and as a ligamentous support to the chain

of ossicles, and a corresponding belief in the unim-
portance of any changes in it which do not invalidate
these functions.   At the same time the true causes of
all symptoms (whether subjective or objective) of
disease in any part of the middle ear, are shown to be
due to morbid conditions behind the membrane rather
than in this structure.

Additional information is given upon the clinical
history of these cases of inflammation of the tympanum
and mastoid cells which terminate fatally, and the
possibility of occasionally averting this end by a timely
perforation of the mastoid process is more strongly
insisted upon.

Two cases of malignant disease of the ear are recorded,
and some of the prominent symptoms in nervous
disorders of the auditory apparatus are attempted to
be explained by considering the intimate relations
which exist between the auditory and the pneumogastric nerves.

London ;
 September, 1880.

PREFACE

FIRST EDITION

AN abstract of these lectures appeared in the 'Lancet' during the latter half of the year 1872, and with some additions and alterations they are now published as originally delivered.

The lectures were given in a conversational manner in the out-patients' department, and as opportunities occurred were illustrated by cases under observation at the time. In republishing them an attempt has been made to describe as shortly and clearly as possible the pathology and symptoms of diseases of the ear, and in directing attention to the treatment of these affections to place before the reader the general results to be expected from remedial measures. Mr Toynbee's book on 'Diseases of the Ear,' and, more recently, Mr Hinton's article in 'Holmes' System of Surgery,' may be considered the two most complete accounts of this subject in English. Whilst fully recognising the immense value of the late Mr Toynbee's researches on the pathology of diseases of each portion of the ear, it must be allowed that since his book was written considerable additions have been made to our knowledge.

This observation is particularly applicable in the instance of non-purulent catarrh of the middle ear and morbid growths within the tympanum, so that in such cases it will be observed the treatment I have recommended is considerably different from that usually pursued by Mr Toynbee. The article by Mr Hinton referred to will be found to contain the opinions of the highest authorities in this branch of surgery.

It is hoped that this course of lectures may take a place intermediate between these two, and so in some measure serve as a text-book for diseases and injuries of the ear.

By permission of the publishers I have made use of five of Mr Toynbee's woodcuts, and I am indebted to my friend and colleague Dr Whipham for the beautifully executed drawings showing the structure of polypi of the ear.

W. B. DALBY.

LONDON;
June, 1873.

CONTENTS

LECTURE I

LECTURE II

LECTURE III

LECTURE IV

LECTURE V

LECTURE VI

LECTURE VII

LECTURE VIII

LECTURE IX

LECTURE X

LECTURE XI

LECTURES

DISEASES AND INJURIES OF THE EAR

LECTURE I

THIS is the first course of lectures that has been delivered at St George's Hospital in the department of surgery under my care, and the subject on which I shall endeavour to give you what information I possess is the practice of surgery in connection with the ear.

So little attention was at one time bestowed by surgeons in this country on diseases of the ear that until Mr Pilcher, Mr Toynbee, Sir Wm Wilde, and a few others devoted their energies to this subject, the treatment of these affections was in a great measure conducted by those who spoke of deafness as if it were a disease, and professed to cure it. I need not remind you that pathology has long since taught us how very exceptional are the instances in which physicians and

1

surgeons may be said to cure disease; still less do
they cure symptoms, and impaired hearing is merely a
symptom of disease. The object of all treatment, so
far as I know, is to relieve symptoms, and to place
patients under circumstances most favorable to recovery.
As we proceed I hope to be able to point out to you
that when that portion of the auditory apparatus which
serves for the conduction of sound is the seat of disease,
it shows under favorable conditions a tendency to recovery similar to what is observed when other parts of
the body are affected, and that when the nervous structures of the ear are subjected to morbid changes we can
at least estimate with tolerable accuracy what progress
and termination may be anticipated for the case.

I shall confine my remarks on the anatomy of the
ear to those points which will be found practically of
use in estimating the changes produced in it by accident or disease, so far as they can be observed by
examination, both on the living and the dead subject,
and so far as they interfere with the function of hearing
and endanger life. Elaborate and lengthy descriptions
can be found in plenty in works on anatomy. At the
same time I must observe that it is necessary for you to
be perfectly familiar with some anatomical facts connected with the form and relations of the external and
middle ear, and these I shall attempt to explain and
demonstrate as we go on.

With regard to the middle and internal ear it is out
of the question to expect to get anything beyond the
most confused ideas on the subject by reading even the
best text-books on anatomy unless the parts have been

frequently seen and handled. If this is done, I think perhaps the clearest and most reliable account of the auditory apparatus in the English language to work with will be found in the last edition of Sharpey and Quain's 'Anatomy.'

The dissection of the internal ear is extremely difficult; but I shall have the opportunity of showing you in the course of the lecture some very beautiful sections made by Mr. Watney, which demonstrate the minute structure of this part. To learn well the anatomy of the tympanum it is only necessary to get a few temporal bones recently sawed out of the subject, and gradually to chip them to pieces with a pair of bone nippers until every part has been seen.

Affections of the ear naturally divide themselves into those of the external, middle, or internal ear. The external ear consists of the auricle and external auditory meatus, and is separated from the middle ear by the tympanic membrane. Although in some of the vertebrate animals the auricle is an eminently useful appendage, it seems open to question, where the loss of one auricle in man through an accident has afforded opportunities of comparing the two ears, whether it contributes in any considerable degree towards collecting the waves of sound in their passage to the meatus.

In the course of the year 1872 we had an opportunity of satisfying ourselves upon this point in the case of a man, W. H—, æt. 21, who came to the hospital in consequence of a wound received in a public-house disturbance. The man with whom he was fighting, after knocking him down, while he was on the ground, bit off his left ear.   Except the lobe which remained the ear was bitten off close to the head.   After the wound had healed, I found, on testing the hearing, that it was not appreciably impaired for sounds proceeding from a point to the left side of the patient, but that the hearing of the right side was slightly the better of the two for sounds which proceeded from either in front of or behind him.   However, the difference was so small as not to be worth consideration.

The external ear is sometimes the seat of malformations, as instanced in an imperfectly formed auricle and meatus, the latter being sometimes so small as only to admit of a probe being passed into it, and occasionally being absent altogether.   Children have been born with four ears, two on either side; and I once saw an adult, in other respects properly formed, in whom the auricles were so small that they measured barely one inch from the tip of the helix to the extremity of the lobe.

A curious deformity of this kind was reported in 1870 by Mr Moxon, of Reading.*   A young girl had in addition to two perfect ears, three rudimentary auricles on the right side and two on the left.

In many of the cases of malformed, or, more strictly speaking, rudimentary ears, which I had seen up to the year 1877, I had noticed a small opening with an oozing discharge, and had considered it to be the opening which led to the external auditory canal.   In November, 1877, Sir James Paget read a paper, which

* 'British Medical Journal' for Nov. 12, 1870.

may be found in the sixty-first volume of the 'Transactions of the Royal Medical and Chirurgical Society,' "On Cases of Branchial Fistulæ in the External Ears." If you will read this paper you will find what had escaped my observation, viz., that these openings are frequently branchial fistulæ, in other words, the openings are " due to incomplete closure of the upper or first post-oral fissure; or, rather, of that part of it which is not utilised in the formation of the Eustachian tube, tympanum, and meatus."

In a rudimentary ear lobe which I removed in 1875 on account of its unsightly appearance, this opening was well marked. It may be seen in the Museum of the College of Surgeons.

Although it is not within the scope of these lectures to discuss the question, I may say that, as the result of observation, I cannot escape the conviction that intermarriages between blood relatives exercise a very considerable influence towards inducing malformations or instances of arrested development in the various parts of the auditory apparatus.

Surgical interference in these cases is not often demanded, and where it has been exercised, especially in instances of rudimentary meatus, the results have not been satisfactory; this being chiefly due to the fact that where the external division of the ear has been defective, the middle or internal one has been so as well.

This is not invariably so, however, and a case has been lately recorded by Dr Morland, of Boston,* in

* 'Transactions of Otological Society of America,' 1870.

which the external auditory meatus of the left side was closed by a cutaneous layer. An aperture was made with the knife, kept open for a time with sponge-tents, and a small gold tube worn in the meatus until the wound had healed. This was followed by considerable and permanent improvement to the hearing. Such examples as this are encouraging, inasmuch as they go towards showing that the parts more deeply situate are not always useless for purposes of hearing. No rule, then, as to the value of operative proceedings can be laid down for these cases, but at any rate an exploratory incision should be made, if any hearing power is present, lest perchance there be more below the surface than might be supposed from the usual results of such attempts at relief.

The length of the external auditory meatus, although

Fig. 1.—The Osseous Meatus Externus of an Infant (*Toynbee*).

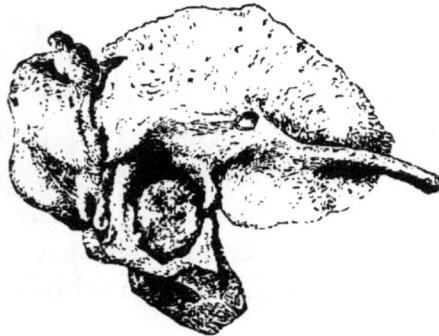

varying in different individuals, is about one inch and a quarter, and the calibre is greater in some than in others. It consists of two parts, the external cartilaginous comprising one third, and the other two thirds being an osseous canal in the temporal bone. At the bottom of this canal is the tympanic membrane, forming an angle of 45 degrees or nearly so with the floor. With infants the osseous meatus is not developed, and the membrane is fixed at the outer part of the skull, being thus nearly horizontal in position. At this time of life the portion which eventually becomes bony is membranous, and later in life the freedom of motion which is permitted between the cartilaginous and osseous part of the meatus when the auricle is raised, is due to some slight remains of the membranous part. Great care then should be exercised in dealing with the meatus of very young children, especially in the case of a foreign body impacted in this situation. In shape the meatus is somewhat oval, the longitudinal diameter at the external part being the longer, but at the other end the transverse. The whole meatus takes a gentle curve forwards. The upper wall of the osseous portion is nearly horizontal; but, proceeding inwards from the middle, the floor dips downwards, and on this account, and from the angle which the tympanic membrane makes with it, it follows that the floor is longer than the roof. It is partly for this reason, and partly because the calibre is smallest about the middle, that a foreign body which has passed this point is so liable to be pressed onwards in the efforts which are sometimes made to extract it; and, again, it is for this reason that, in using the speculum, it must be tilted a little in order to obtain a complete view of the membrane.

FIG. 2.—VERTICAL SECTION OF EXTERNAL AUDITORY MEATUS.

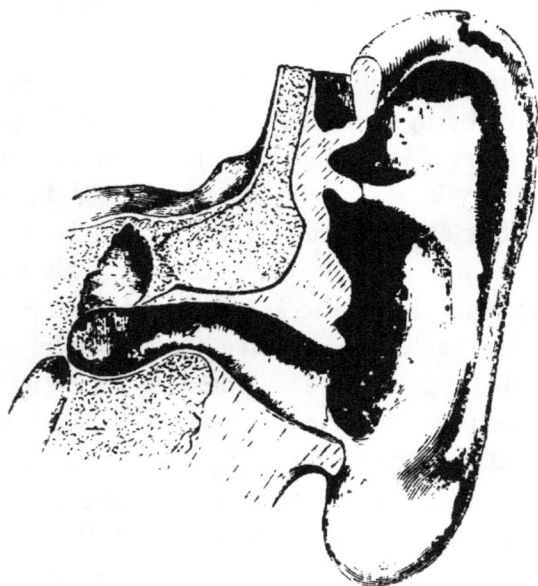

The meatus is lined throughout with skin, which is continuous with that of the auricle, and a thin layer of it is prolonged over the tympanic membrane.

The method of examining the external auditory meatus and the tympanic membrane, which you will see practised here, and which is now almost universally employed in Germany and in this country, was originally introduced a few years ago by Dr von Tröltsch, the Professor of Aural Surgery in Wurtzburg. It seems almost strange that so simple and effective a plan had not been thought of before, in place of the imperfect methods previously in use.

If it were necessary to examine the interior of any curved tube closed at one end, the best way by which to obtain a view would be—1stly, to straighten the tube as much as possible; and, 2ndly, to illuminate the interior. For some time in the case of the ear the first of these conditions was fulfilled by introducing a straight tube into the meatus; and the second, by taking the patient to a window and turning the side to be examined towards the light.

It was found that a patient was not always moved with facility into a convenient position, and that when the light passed into the speculum the surgeon's head was often in the way as he attempted to look down it. The next improvement was Miller's lamp, by the aid of which light from a candle was thrown into the speculum from a steel reflector.

The plan referred to as being practised in the present day is at the same time simple and effective. It consists in reflecting light from a concave perforated mirror down a funnel-shaped speculum of the kind first known as Gruber's. In consequence of the variety in the size of the meatus, several specula should be at hand (perhaps half a dozen), the straight tubular part being oval in shape, and having varying calibres. The straight portion is generally made too long; it should never measure more than half an inch, and the whole length of the speculum should not exceed an inch and a half. If longer than this, not only has the light to be sent down a greater distance, and therefore the

reflector be held farther off, not so good a view being
in this way obtained, for the eye is thus placed so
much farther from the object, but it is found to be in
the way when an instrument is being used, as in
removing a polypus. It is unimportant whether the
speculum be made of silver or vulcanite : the latter
kind, perhaps from the contrast in colour, renders
slight changes in this respect on the tympanic membrane more strongly marked. Both kinds remain in
the ear without holding. With regard to the specula
which are made to dilate with a screw, and the speculum of Kramer, which opens out with handles after it
has been introduced, it is idle to suppose that the
meatus under ordinary circumstances will admit of
dilatation, and therefore for general occasions they are
useless. There are some cases where Kramer's speculum may be serviceable, and these are those in which
the tragus is situated in a more advanced position than
natural, and the anterior and posterior boundaries of
the cartilaginous part approximated more than usual,
giving rise to a narrowing of this part, so that the orifice is more like a slit than the ordinary shape. These
are, however, exceptional instances ; and, speaking
generally, the object in using a speculum is, not to
dilate, but to make a straight tunnel for light, and as
the tunnel can be moved about, a complete view of the
meatus and the tympanic membrane can in this way be
obtained.

In introducing a speculum into the ear the auricle
should be pulled upwards by the left hand, as this will
raise the external movable part of the meatus to a level

FIG. 3.—HAND MIRROR.

$\frac{1}{2}$

FIG. 4.—EAR SPECULA.

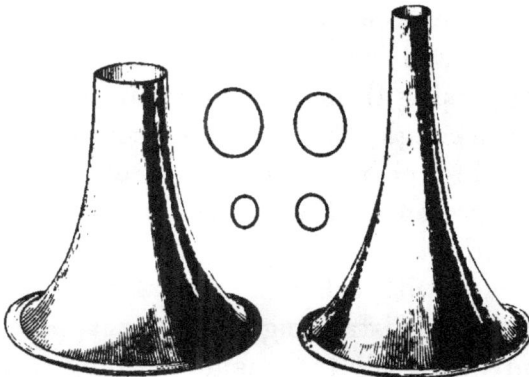

with the rest of it.   I call your attention to the necessity of exercising great gentleness in using the ear
speculum, but especially on the first occasion of seeing
a patient, for even when the meatus is not affected in
any way some persons are peculiarly sensitive in this
part.   I have on many occasions seen a feeling of
faintness to be induced during an examination, and
twice in the course of the last three years the patient
has positively fainted.

You will find that on examining a large number of
cases some persons will be found in whom an uncon-
trollable cough is induced by the presence of the
speculum.   This is due to reflex action excited by
irritation of the auricular branch of the pneumogastric nerve, and this personal peculiarity, so to
speak, depends upon a more than usually superficial
distribution of this branch in the meatus.   I have
met with a great many cases in which some local irritation in the external auditory canal have been the
cause of cough, which has persisted for years, and which
has ceased after the removal of the local trouble.   This
has happened especially when there has been a polypus
in the canal ; and sometimes the irritating influence of
discharge passing from the tympanum over the canal
will induce a cough; the fact of the cough and the discharge ceasing or returning synchronously with the
absence or presence of the irritation will point to the
cause and effect.

In common with others I have also observed many
cases where a distressing cough has persisted for
years and defied all treatment, until by some happy

circumstance attention has been called to the ear, the foreign body removed, and the symptom of cough has disappeared.

In simple examination the reflector is held in the right hand, and when it is desirable for both hands to be free, as in any operative proceeding, it is worn on the forehead and fastened round the head with a band in the same way as in examining the throat with a laryngoscope. Thus two kinds of reflectors should be

Fig. 5.—Mirror with Frontal Band.

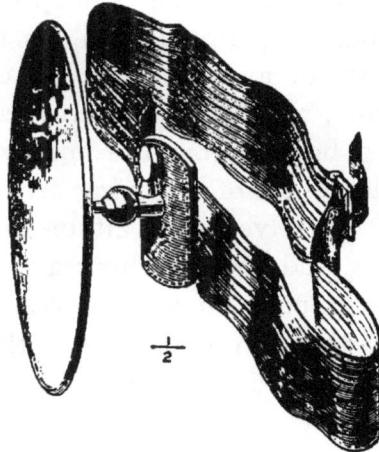

at hand. The best light for minute examination of any object is a bright diffused daylight, and this applies as much to the external auditory meatus and tympanic membrane as to any other. The patient should either be seated or stand close up to a window, with the head inclined a little downwards, and the side to be examined turned away from the light, but somewhat sideways. If the day is too dark for a good view, as unfortunately it often is in this country, the light may be reflected from a bull's-eye lamp, lit with gas, fixed to a stand, and arranged that it can be moved up and down so as to be placed on a level with the patient's ear when he or she is seated near. This artificial light is not nearly so good as daylight, as, from the translucency of the membrane, a somewhat yellow tinge is imparted to this structure which is not natural to it. If the patient be in bed, a moderator lamp is a convenient source from which to obtain light. In the absence of this a Miller's lamp may be used.

Having become acquainted with a means by which the external meatus and tympanic membrane can be readily examined, we shall not at any rate attempt to extract foreign bodies from the meatus, or syringe the ear with the object of removing cerumen, until we have satisfied ourselves that either the one or the other is present, and though this may appear a needless caution to impose upon ourselves, and not very complimentary to our intelligence, I think it necessary to mention it, inasmuch as it is not a very unusual occurrence in the out-patients' room for children to present themselves whose ears have been sedulously probed to ascertain whether the story given of something being in the ear is true, and adults whose ears have been perseveringly syringed in order to relieve them of cerumen, when the meatus has been quite free from this secretion; the treatment pursued in either case not proving useful, and sometimes (as you will have an opportunity of seeing if you attend in this room) very much the reverse.

Foreign bodies in the meatus are generally put there by children at play, and are chiefly stones, beads, peas, and the like.

You know that it is an unvarying rule in surgery, that before a patient be cut for stone, while he is on the operating table the presence of the stone should be unquestionably demonstrated. It ought to be a no less unvarying rule in the case of a foreign body in the ear that its presence should be demonstrated by sight before any attempts are made to extract it, and more than this, the operator should have a full view of what he is doing during the whole time he is endeavouring to extract it. You may accept this unreservedly as a maxim never to be departed from. You will recollect a few weeks ago two cases which bear upon these points, one in which a small stone had been thrust into the ear of a little boy. The stone, which we subsequently saw, was somewhat larger than a pea, and had a smooth surface. The attempts which had been made to remove it had unfortunately resulted in pushing it further in, and when the child came to this clinic, the tympanic membrane had been ruptured, and the stone was lying (we could see it and touch it with a probe) in the cavity of the tympanum. There was a profuse discharge from the ear, and a considerable amount of swelling of the meatus. Although I could see the stone, I could not, in consequence of its smooth and slippery surface, grasp it with the forceps, and even if I could have done so, as there was only just room for it to pass the narrowest part of meatus, I should not have been able to withdraw the forceps, as they enclosed the stone. So nothing was done, except that the ear was gently syringed every day, and in three weeks, the swelling of the meatus having subsided, the stone dropped out. The other case was one in which a bead was in the meatus and was removed without any difficulty. Please, therefore, in similar cases bear in mind these two points.

In the first place, as the meatus is lined by skin continuous with that of the auricle, provided the edges of the objects are not sharp, they may remain there for an indefinite period and do no harm. In the second place, very considerable injury is often done by injudicious efforts which are made to extract them.

I cannot whilst on this subject help endeavouring to impress on you the importance of care in these cases, and reminding you that on two successive days at this hospital I had the unhappiness of seeing four instances in which the tympanic membrane had been ruptured by attempts of this kind, and in two of these an examination proved that there had been no foreign body in the ear. I may also tell you that not so very long ago I had my attention directed to a still more striking example in which the patient died during attempts to extract from the ear what a post-mortem examination proved had only existed in the conjoint imagination of the parents of the child and the operator.

When you have seen for yourselves foreign bodies in the ear, do not, however, understand me to say that no attempt should be made to remove them. But no instruments should be used, excepting when, with the mirror on his forehead, light is reflected down the meatus, and the operator, with both hands free, can see what he is doing.

With this precaution you will have no difficulty in extracting any foreign body from the ear, if it is of such a form that it can be closed upon by the ring polypus forceps, or is one which presents such a surface that the forceps can get a hold upon, that is, when it is of a

FIG. 6.—RING POLYPUS FORCEPS.

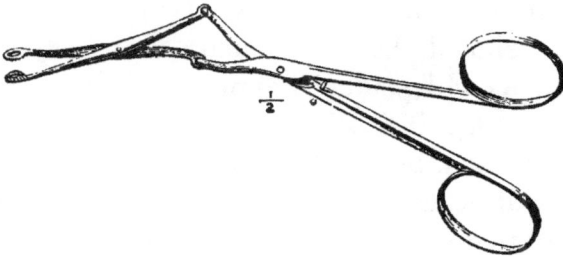

soft nature or uneven shape. A loop of silver wire may sometimes be passed beyond it, and by pulling at this it may be moved outwards; or the noose of a Wilde's snare may be used in the same way, and the body sometimes by this means be secured and extracted. I once removed a cherry-stone in this way from a child's ear. Young children must have chloroform, not because they are being hurt, but because they will not keep quiet; and without this it is impossible to do any good. Adults will sit still in a chair, as they are not put to any pain. A very useful and simple little instrument, and one which I have used constantly for the last ten years, is a small piece of steel, not

2

thicker than a fine wire, and slightly bent at one
end so as to form a very minute hook; the other
end is fixed into a handle. With this little hook the
object may be lightly touched, turned round, and
drawn towards the external opening of the canal. If
the foreign body cannot be removed in one of these
ways, or if it have passed more than half way down,
for the reason shown when speaking of the shape of
the meatus, any further attempts will assuredly only
have the effect of sending it further in, and will frequently result in a rupture of the tympanic membrane.
The ear should be occasionally syringed, and the
foreign body will gradually approach nearer and nearer
the orifice, until it finally drops out. It may be weeks,
or even months, before this happens, but if it be left
alone it will do no harm. If the tympanic membrane
be unfortunately ruptured, this is generally followed
by suppuration in the cavity of the tympanum, a discharge from the ear lasting a long time, and more or
less permanent deafness. After such an accident, if
there be much swelling of the meatus and pain, a few
leeches in front of the tragus, followed by fomentations, will be advisable: and when by-and-bye the
offending body has come away, the case must be
treated in the way we shall consider in a future lecture
on perforations of the tympanic membrane. A most
clever method for removing foreign bodies from the
ear has been suggested and successfully practised by
Dr Löwenberg of Paris. The end of a rod is dipped
into melted glue; the point thus armed is held in contact with the foreign body until the glue has hardened

(about twenty minutes suffices for this); the rod is then withdrawn, bringing away with it the foreign body.

Anything more energetic than the treatment I suggest in these cases I believe not only to be unnecessary but harmful; and I must confess to looking with disfavour on any of the numerous instruments which have been brought forward, it is said, with the special object of getting behind the foreign body and removing it, for I have never met with any cases where I think they would succeed when the more simple means fail; and, on the other hand, I have frequently seen injury to follow their employment. However, do

FIG. 7.—RECTANGULAR FORCEPS (*Toynbee*).

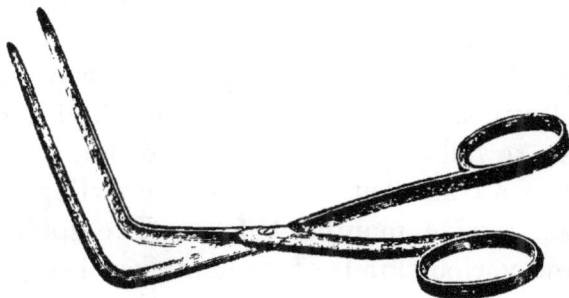

not let us do an injustice to ingenious contrivances we have not seen, which may some day, possibly, help us to accomplish what is often a very troublesome task. Any sort of instruments or forceps, however, which may suggest themselves to you at the time may be employed (ordinary rectangular ones I have often found useful) as you cannot do any harm, so long as you never forget that it is not only silly but mischievous to trust to touch instead of sight, and to fish about in the dark when you have an easy means of illuminating the meatus at the time.

Mothers often bring their children and ask to have various things taken out of their ears, and on examination it is found that there is nothing in them at all. In the case of adults, it is quite surprising what mental distress they appear to suffer from the notion that there is anything in the ear, and it is occasionally, when there is nothing there, very difficult to persuade them of their mistake.   If an insect should crawl into the ear it gives rise to most unpleasant sensations; the patient and his friends generally resort to a variety of expedients to kill it, tobacco smoke, &c.   A few drops of oil or water poured into the ear will either kill the insect or make it creep out, or at any rate a little gentle use of the syringe will always bring it away.   I remember once having seen a highly nervous condition to follow a case of this kind in a young lady, who gave an account of a spider crawling into her right ear twelve months before.   The spider, after many and various kinds of attempts had been made to get it out, was at last, so she said, extracted with forceps, but not before it had remained in the ear for four days.   Ever since then she had suffered from painfully acute hearing; the sound of her voice was disagreeable to her, and after every sound she continued, for some seconds, to have a ringing in the ear.

Among the remarkable things met with in the meatus, I may mention that Dr J. Green, of St Louis, at a meeting of the Otological Society of America in 1870, related a case of aspergillus, which from the history would seem to have remained in the ear for two years; and Dr Blake, of Boston, two years later, reported two cases of living larvæ in the ear.

With regard to cerumen in the ear, the first symptoms of its presence is generally a loss of hearing power. This symptom sometimes comes on gradually on account of increased secretion until the meatus becomes blocked up with it, or it may, which is the more general way, come on suddenly. And this occurs in the following way:—The cerumen from being secreted too abundantly on the walls of the meatus gradually narrows the passage by which sonorous vibrations proceed to the tympanic membrane. By a change of position of some portion of the cerumen, caused either by water getting into the ear, the patient putting his finger into the meatus, picking the ear, or using the twisted end of a towel to clean the ear (this bad practice is often followed by nurses with children), the already narrowed passage becomes altogether closed. Some other movement in like manner may for the time bring back the hearing by restoring the passage. Another symptom often evoked by cerumen in the meatus is tinnitus, and it is sometimes of the most distressing character. Anything which causes undue pressure of the stapes on to the fenestra ovalis, and so on the labyrinth, will produce it, and in this case it is brought about by a hard plug of wax directly in apposition with the tympanic membrane, thus transmitting pressure through the malleus and incus on to the stapes. The same exciting cause will often induce attacks of giddiness, and I have known the symptom to have existed even for years without eliciting suspicion of its cause, and the patient to have been submitted to all sorts of treatment for its relief, without of course receiving any benefit. This arose from the fact that there was no accompanying deafness to point to a cause, for in these cases the hard plug of cerumen does not always entirely cover the membrane, and so there is ample room for sonorous vibrations to fall upon it and thus be transmitted as before. As a rule, however, there is deafness as well.

One case I was very much struck with some years ago, where an old lady had had rather alarming attacks of giddiness at various times for more than three years, and not until she had become deaf (I suppose from a little more cerumen completely covering the membrane) had attention been drawn to the ear. The giddiness and deafness both disappeared with the removal of the wax.

Accumulations of hardened cerumen from continued pressure on the membrane will occasionally produce ulceration of this structure, thus becoming the direct cause of a perforation, and in the same way they have occasionally induced partial absorption of the bony walls of the meatus.*

The only legitimate way of removing cerumen is by the use of the syringe, and the best form is one where the nozzle is very small and can be removed from the other part by bayonet action. In using this kind, time is saved, as the syringe is quickly filled, and from

* There are some preparations showing this in the Museum of the College of Surgeons.

the small size of the nozzle the meatus is not blocked up as it is in most of the syringes made. It is only necessary to use warm water of a pleasant temperature. In so simple an operation as syringing the ears there is a skilful and an unskilful method of doing it; the first is of course the more effective one, and it is from not taking the trouble to learn it that some persons are so long in getting out cerumen which a few syringe-fuls of water properly applied will at once expel. The auricle should be drawn upwards between two fingers of the left hand so as to put the whole meatus in a straight line, and the nozzle of the syringe should be kept in close contact with the roof of the meatus. In ordinary cases the mass will be easily dislodged, but if it has lain there for a long time, perhaps several years, and become as it will do sometimes of almost stony hardness, its removal is not so easily effected. In such a case the patient should lie on the opposite side for a time and have the ear filled with water so as to let the cerumen soak and become softened. It will, however, occasionally be so hard as to render it necessary to pour in water or oil (it does not matter which) for two or three nights successively. It is well not to use the syringe too long at a time; and at intervals, during the proceeding, the ear should be examined with the speculum to see if the canal is clear, for if it is so, considerable irritation may be set up by syringing directly on the membrane; indeed, it is not very rare to meet with patients (I have seen several) whose ears have been syringed vigorously when there was nothing to bring away, and they have had inflammamation of the membrane and subsequently a perforation. The wholesale way, therefore, in which people get their ears syringed as soon as they feel deaf is not altogether without its drawbacks, and there are some points that are worth attention even in cases of wax in the ear. When the cartilaginous part of the meatus has become narrowed to a mere slit, by approximation of the anterior and posterior boundaries, it is sometimes extremely difficult to get away cerumen, and I remember one old woman (and by the way, a meatus of this kind is generally met with in elderly persons, I cannot say why) who had to come on five different occasions to the out-patient room before the secretion was entirely removed. This alteration from the natural shape of the meatus will sometimes amount to almost complete closure and then become the cause of deafness. In these cases assistance may be given to the hearing by wearing a small silver tube which will keep the opposed sides apart.

LECTURE II

IF I attempted to describe to you all the affections
of the external auditory meatus that you will find
classified in some of the text-books on aural surgery,
I should not only be confusing you, but I should, in
their recital, become very confused myself. For example, I fail altogether to understand what the late
Mr Toynbee meant when he described a number of
cases as "simple chronic inflammation of the dermoid
meatus," "chronic catarrhal inflammation of the dermoid meatus," and "catarrhal inflammation of the
dermoid layer of the external meatus, with caries of
the posterior wall."   To say the truth, I do not know
what is meant by catarrhal inflammation, except as
applied to mucous membrane, and can only tell you
that the external auditory canal is subject to inflammation; that, at one time, the integument is alone
affected, at another, the periosteum and bone are
involved; and I will try to describe to you the way in
which you will commonly find patients suffer, and the
most convenient mode of relieving them.   1stly, then,
if the area of inflammation be strictly circumscribed,
there will be boils and abscesses in the meatus.
2ndly, if such an area is not circumscribed, a larger

and less definable part of the ear will be affected, and the term " diffused " may be applied to it.

Boils in the meatus give the subjects of them a great deal of pain and trouble. If they are situated near the external orifice, they are not nearly so painful as when they are somewhat farther in the meatus. In the latter case, you will find it almost always advisable to open them, as the sufferings of the patient are thus very much shortened. The local treatment is sufficiently simple, and may be said to be as follows :— Fomentations until the boil is ready to be opened, or the boil discharges itself. But this is only a small part of the treatment. The difficult and most essential matter is to change the condition of health or the habit of life which induces the boils, for they seldom occur singly. Generally, a few days after one is well, another will form either in the same ear or the other one, and so on sometimes for weeks or months. It is also a rather singular fact that patients with this affection scarcely ever have boils in any other part of the body. I cannot say either that they are confined to persons who may be said to be unhealthy or weakly.

The most obstinate case I ever saw occurred in a young man, an exceptionally strong person, who spent the best part of his time in field sports and athletic exercises, lived almost by rule, and was in training generally two or three times in the course of the year. For three years he had never been one week quite free from a boil in either ear. With cases of this kind you may be quite sure that there is some habit of life which wants correcting ; you must find this out if you want to give permanent relief to your patient, and it will sometimes tax all your ingenuity to do it. In the instance I mention, the patient was accustomed to drink beer—about one pint in the day; this was left off, and no stimulants allowed beyond a little claret. Small doses of Carlsbad water were taken every day, and he got quite well; thus, apparently very slight change in his diet was made, but it was quite enough to free him from the inconvenience he had so long suffered from. Another case, perhaps, will require entirely different management; in short, constitutional treatment, a term which includes so much. Exercise in the fresh air, stimulants for those who require them, an absence of stimulants for those who take too much, appropriate diet, and medicines to suit each case.

Cases of abscess in the meatus are sometimes very troublesome, and require a good deal of care and management. The first symptom is pain in the ear, which in a few hours becomes so acute as to put sleep out of the question. After twenty-four hours or so the meatus in its entire extent will be swollen; sometimes the whole external ear will be enlarged, red, and tender to the touch. The movements of the jaw occasion great pain. There is more or less deafness in proportion to swelling. If the patient be seen within a day or two of the commencement of the attack, the greatest benefit may be afforded by two or three leeches placed in front of the tragus, just over the articulation of the lower jaw, and followed by fomentations: in this way the affection may be cut short occasionally. When the leeches are applied, the meatus should be filled with cotton wool, to prevent the blood getting in. If not checked in the early stage, the state of things described will increase in severity for several days, and, on examining with a probe, one part will be felt to be far more acutely tender than any other part of the meatus. The whole meatus is so swollen that it is not possible to see the position of the abscess with a speculum. This point having been distinctly localised, the sooner an incision is made into it the better, as by this the tension of the parts, to which the agonising pain is due, is relieved. The best knife to use is a curved sharp-pointed bistoury, and a free incision should be made as it is withdrawn. After this the patient gets well in a few days. Very frequently, however, he is no sooner well of one abscess than another commences to form, and all the suffering has to be endured again. It is this recurrence that makes the affection such a troublesome one. Sometimes six or seven abscesses will occur successively in the same patient, obliging him to remain in the house for as many weeks, and causing an amount of suffering altogether disproportionate to the gravity of the illness. In recognising this occasional peculiarity in these cases, you will do well to observe caution in promising a patient that he or she will be well within a certain time ; and you will warn them of the possibility of a recurrence of the abscesses, for unless this be done they will be very apt to think that the incision made for the relief of the first abscess was imperfectly formed.

The diffused form of inflammation of the external

auditory meatus differs from the circumscribed, inasmuch as it does not terminate in abscess, and, as its name implies, in being diffused throughout the integument of the canal.   Since the outer cuticular layer of the tympanic membrane is continuous with that lining the meatus, there is always a certain amount of risk lest this membrane become implicated, and for this reason this affection is more serious than the one just considered.   It is not strictly correct to describe it as being of two kinds—viz. acute and chronic,—because, although the attacks are acute at one time and chronic at another, their gradations so insensibly pass into one another that an artificial division is practically useless. Children are especially liable to it; and as it is generally neglected among the poorer clasees, with them it is not unfrequently the origin of a perforation of the tympanic membrane.   It often succeeds the exanthemata, but not nearly so commonly as does purulent catarrh of the middle ear spreading along the mucous membrane from the throat.   The diffused form of inflammation of the external meatus commences with a feeling of uneasiness and itching just within the orifice which very soon becomes red, tender, and swollen ; and there is pain during mastication.   The whole ear is red and swollen, and has the appearance of standing out (so to speak) from the head.   These symptoms increase in severity until, with the appearance of a discharge, the pain ceases, and the swelling gradually subsides.

Such is briefly an account of its acute stage, which usually occupies a week or ten days.   The treatment is palliative—viz. leeches and fomentations ; but as this affection does not occur to persons in a good state of health, the diet should be attended to, and a change of air, if possible, be resorted to, for it is most desirable not to let this condition become a chronic one, as it is most apt to do.

After the appearance of the discharge, which is generally not very copious, the meatus should be kept carefully cleansed by syringing every day, and afterwards a mild astringent lotion may be used. After an attack of this kind, as the external layer of the tympanic membrane has shared in the general condition of the lining of the meatus, the natural translucency of the membrane is lost, but this in no way interferes with its functions.

Injuries to, and morbid growths of the auricle present no especial points of interest to distinguish them from the same conditions elsewhere. This will also apply to skin diseases in this situation. Eczema, perhaps, as being a common affection of the outer ear, calls for a few words of notice. It may be acute or chronic. After repeated attacks (for a patient who has once suffered from it will be very liable to it again) the auricle becomes very much thickened, and the entire meatus shares in the same condition. If there is any deafness, it is only so far as may be accounted for by the narrowing of the canal, which takes place when the disease has lasted for a long time. Like eczema in any other part, it is very obstinate; but if the patients will consent to keep to rules of diet, including an absence of stimulants, they will get well. This is more important than any external applications.

Of these, very mild mercurial ointments, are, perhaps, the most useful; and before they are applied the parts must be carefully dried.

Exostoses are not unfrequently met with in the external auditory meatus, and are not confined to any situation in the course of the osseous part of the canal. They are solid bony growths, covered with periosteum. Sometimes there are two or three at one time, and they scarcely ever have a pedicle. On this account, and from their position, they are not easily removed by an operation. Fortunately, this is not often necessary, as they seldom entirely close the meatus, and so do not interfere very materially by their presence with the hearing, but do so indirectly by secretion of cerumen and epidermis collecting behind them; and in this way may be very troublesome, for, besides acting as a mechanical obstruction, the constant pressure of hardened secretion upon the tympanic membrane may induce ulceration of this part.

The ear will in each case require the most careful syringing from time to time, and it is sometimes extremely difficult to get the point of the syringe between the growth and the wall of the meatus. Moreover, it is not always a desirable proceeding to syringe the ear, inasmuch as there is a difficulty in drying the canal completely, and if a little water is left behind the growth it may become the source of irritation; any secretion may therefore be better removed by forceps if possible. On examining an ear the external canal may be found to be nearly closed by the meeting of three growths, and the other ear, perhaps, will

represent the exact counterpart of its fellow, no history
of any kind giving a clue to the origin of such changes.
Examples of this kind are probably congenital. On
the other hand, however, when one ear alone is affected,
a history of a discharge from the ear, dating, perhaps,
from many previously, may often be elicited. I am
very much inclined to think that the exciting cause of
the growth is the passage of the irritating discharge
along the canal. Dr Roosa, of New York, seems to be
of this opinion too in the account of five cases which he
has recorded in his translation of Dr von Tröltsch's
book on the ear; he there gives a very good chapter
on this affection, and says that these bony formations
should be considered " rather as general enlargements
of the periosteum and bone structure immediately
beneath, than tumours, true exostoses " (a view I entirely agree with). Mr Toynbee considered them to be
a result of gouty or rheumatic diathesis, and Dr Gruber
has noticed in syphilitic patients a general enlargement
of that part of the temporal bone which forms the
external canal. When, however, the enlargement is
only in the one external canal, Virchow's view in
seeking for a local impression as an origin of such
hyperostoses, appears to me more rational than putting
it down to any constitutional affection. This local
impression may take the form of discharge passing
over the canal, or the repeated pressure of boils or
small abscesses. No doubt these growths may remain,
and do remain for long periods without attracting
notice, until some slight accumulation, by completing
the closure of the canal, calls attention to their exist-

ence.   If this secretion can be taken away it is better
that no more severe measures be adopted, but it
occasionally becomes imperative that they should be
removed.   For example, if behind the bony enlargement there is a perforation of the membrane, and the
discharge cannot pass outwards, or if from the tympanic cavity a polypus arises, in either of these cases
the life of the patient may be in jeopardy.

To remove an exostosis in the auditory canal is no
easy matter.   The operation must be done under
reflected light, and the patient's head, as well as the
surgeon's, must be absolutely still.   Of course, an
anæsthetic must be given.   The bone is generally of
ivory-like hardness.   In 1876* I gave an account of
what I found to be the most practicable method of
operating, and I have since then found no better plan.
It consists in drilling away the bone by means of a
drill in common use amongst dentists.   Some time
previously I had employed a plan which was successfully practised and originally suggested by Dr F. P.
Clark, of Bristol, in 1873.   This consisted in passing
a continuous current into the growth through needles
united at its base.   I may say that in using the drill
an assistant must turn the lathe and cease turning
according to the directions given by the operator.
The proceeding is often most tedious, as the bone
sometimes bleeds freely, and has to be constantly
dried each time the drill is applied.

When syphilis shows itself in the outer ear, it does
so as condylomata in the meatus, or, more correctly

* 'Lancet.'

speaking, mucous tubercles, or secondary eruptions on the auricle. These call for no especial treatment beyond what is necessary for similar conditions when they are met with in other parts. The manner in which syphilis affects the middle and internal ear will be considered by-and-by.

The external ear is occasionally the seat of malignant disease. Sir W. Wilde mentions two fatal cases, one in a woman, aged 50, where death occurred three weeks after the appearance of the fungous growth, and another in a boy of seven, in whom the cancerous growth had affected the petrous and mastoid portions of the temporal bone. It would seem that generally when malignant disease is situate in the ear, it commences in the mucous membrane lining the tympanum. Mr Toynbee relates three cases illustrative of this: in two cases which came under my notice in the year 1878, and which I shall mention when I come to speak of diseases of the middle ear, the seat of origin was undoubtedly the cavity of the tympanum.

I have no more to say about the external ear, and as in the next lecture we shall begin the subject of diseases of the middle ear, let us devote a few minutes to the anatomy of it.

To make a dissection of the tympanum, it is well to use a temporal bone as soon as possible after its removal from the subject, as the muscles very soon get dry, and their action cannot be seen on dragging on them with forceps. The whole of the osseous part of the meatus should be chipped away with a pair of bone cutters, the roof of the tympanum picked off with forceps, and the tympanic membrane carefully cut away from its attachments. In the dissections before us made in this way the cavity of the tympanum is well seen, with its contents.

Looking from without, straight into the cavity, it will be seen that the chain of ossicles is not in the middle line of the tympanum, but is placed nearer to the roof. The first object on the inner wall that strikes the eye is the promontory, the shape of which is expressed by its name. It is a rounded prominence of bone corresponding to the first turn of the cochlea, and in the dry preparation is seen to be marked with grooves for the tympanic plexus of nerves; at its posterior part is a large hole, the fenestra rotunda which opens into the cochlea, and is separated from this in the recent state by a membranous septum. Above the promontory is seen the stapes, the stirrup-shaped bone, the base of which marks the position of the fenestra ovalis. The longest diameter of this fenestra is from before backwards, and it is, like the other, separated by membrane from the labyrinth. It opens into the vestibule.

The chain of ossicles and the fenestra ovalis therefore are situated above the promontory, and again above them is a ridge marking the position of the aqueduct of Fallopius, in which is contained the portio dura. Higher still is the roof of the tympanum; this is a very thin plate of bone (by the way it is much thinner in some subjects than in others) which separates the cavity of the tympanum from that of the cranium. This relation it is of great importance to bear in mind, as will be seen when we come to consider disease of the tympanum spreading to the brain. Notice also the relations to the tympanum of the lateral sinus, the jugular fossa, the carotid artery, and the mastoid cells. These cells, you see, communicate very freely with the cavity of the tympanum.

A little behind the stapes on the posterior wall is a nipple-like process, whose apex presents a foramen. This object is called the pyramid, and is the bony canal which protects the stapedius muscle. I call your attention particularly to this muscle as it emerges from its canal, and the direction a force would take if acting in a line with the muscle which by its other extremity is attached to the head of the stapes. If two bristles are passed one on either side of the processus cochleariformis they will be observed to enter at the anterior wall of the tympanum. The upper one indicates the position of the tensor tympani muscle, whose direction at first is right across the tympanum (both the tympanic muscles are thus protected by bone): the lower one is the Eustachian tube. The passages leading out of the tympanum are, you observe, considerably above the floor of the cavity.

The processus gracilis of the malleus is fixed into the Glasserian fissure at the anterior part of the tympanum; the short process of the incus into the mastoid cells at the posterior part of the tympanum. A line drawn through the two will represent the axis around which the chain of bones rotate, so that any rotation from without inwards will press in the stapes and at the same time drag inwards the tympanic membrane;

while any rotation in the reverse direction will drag on the stapes and press the tympanic membrane out- wards. Rotation, however, in this latter direction as affected by muscular force, or, to speak correctly, muscular relaxation, is confined to simply a return of the tympanic membrane and ossicles to their original position, as will be evident if we consider the attach- ment of the muscles and their action. Rotation from without inwards is effected by contraction of the tensor tympani muscle, whose position as it enters the tym- panum has been seen to be at first horizontally back- wards. At the termination of the canal it takes a turn outwards, and sliding on the edge of the processus cochleariformis is inserted into the malleus at that point where the handle joins the neck of the bone. This point lies below the axis around which the chain of bones rotate. From this it results that the mode of action of the muscle is as follows :

Upon its contraction the handle of the malleus is drawn inwards, and if motion in this direction could be continued, the extremity of the handle of the malleus would at last touch the promontory. The leverage obtained by the muscle in its action is, how- ever, not very great, as the insertion of the muscle is so little below the axis or fulcrum. When the malleus is drawn inwards the tympanic membrane goes with it ; this movement at the same time rotates the incus, and thus the stapes will be pressed on to the fenestra ovalis. But as the movements of the stapes in this direction are necessarily of a very limited character, any contraction of the tensor tympani which continues after this required pressure is produced has the effect of rotating the malleus on the incus. This further action of the muscle (so to speak) continues to draw in the tympanic membrane, so that there is proportionately more movement admissible for the membrane than for the stapes.

It seems to be pretty generally agreed among physiologists that the stapedius muscle affects the voluntary act of listening. It appears to me probable that the contraction of the muscle produces a tilting movement of the stapes, with the result of pressing the posterior part of the bone into the fenestra ovalis, and lifting away the anterior part. If this be so, whatever pressure on the labyrinth may be taken away by the withdrawal of the anterior part is compensated for by the inward movement of the posterior part. Thus, the effect of each contraction of the muscle cannot be to remove pressure from the fenestra ovalis any more than to induce it; at any rate, the movement of the stapes must of necessity induce a corresponding movement in the fluid contents of the labyrinth. May it not be possible that this general motion imparted to the fluid in the labyrinth should make the nerve more sensitive to sound? An ingenious suggestion was made to me by Mr Stewart, the curator of the Museum of St Thomas's, with regard to this question. He thinks it likely that the change in position of the base of the stapes, which follows contraction of the muscle, may cause the waves of sound to have an increased effect upon the anterior part of the labyrinth, in which, as you know, the cochlea is placed. You will find in

works on physiology and anatomy many other explana-
tions on this point, which, at present, does not seem to
be settled among physiologists. Until muscular fibre
has been demonstrated in the structure termed laxator
tympani you may regard it as a ligament.

The bony portion of the Eustachian tube (about
three quarters of an inch in length) will just admit a
small probe. The cartilaginous part (about one inch
long) increases in calibre until it reaches the pharynx;
you can see this opening in a vertical section of the
head (a moist preparation) on a level with and just
behind the inferior meatus of the nose. The chief
points to remember about this tube, which with the
tympanum makes up the middle ear, are, that it is in
life generally closed, that during the act of swallowing
the sides of the faucial opening are drawn apart by
the palate muscles, that its lining mucous membrane,
beginning from the pharynx, gradually becomes more
delicate, until, as it lines the tympanum, it is extremely fine in structure, is closely attached to the walls
of this cavity, invests the ossicles, covers the fenestra
rotunda, lines the mastoid cells, and forms the inner
layer of the tympanic membrane.

The tympanic membrane has for an external layer
a continuation of the skin which lines the meatus.
Between these two comes the membrane proper, composed of outer radiating fibres derived from the
periosteal lining of the external meatus, and inner
circular fibres from the periosteum of the tympanum.
The curvature of the membrane is not, strictly
speaking, a concavity, but rather a tucking in; for

the parts not dragged upon by the handle of the malleus show an inclination rather the other way.

Very soon after death the membrane loses its lustre, so that examinations conducted with a view of contrasting healthy with diseased appearances must be made on the living. As you reflect light from this mirror down the speculum and look through the hole in the mirror, you will see the tympanic membrane at the bottom of the auditory canal. It is, you observe, translucent, lustrous, and of a bright slate colour. At the upper part is the short process of the malleus shining through and pressing on the membrane. Passing down the middle from above is the handle of the malleus, terminating in what is called the umbo. From this downwards and forwards is seen what is spoken of as the bright spot. This appearance, observed on looking down a speculum with reflected light, is due, according to Politzer, to the inclination given to the membrane by the traction of the malleus; for the light, being cast upon the membrane, is for the most part reflected on to the lower part of the meatus, but on the spot signified is reflected directly to the eye of the observer. A wave of sound falling upon the tympanic membrane will set in motion the air within the tympanum, and these vibrations will affect the membrane of the fenestra rotunda. Furthermore, there are the vibrations which act upon the chain of ossicles. These bones are so arranged as to make their vibrations, as a whole, so much more effective than those of their particles, that these subsidiary vibrations may be put out of consideration. The fluid contents of the

labyrinth being thus set in motion, the terminations of
the auditory nerve are excited; in other words, hearing is effected.   Whatever circumstances, therefore,
interfere with the proper performance of the functions
of this conducting apparatus (and in it must be included the Eustachian tube, whose office appears to be,
by admitting air from the pharynx, to ensure the same
pressure from air within the tympanum as from the
air external to the membrane) will come under the
head of diseases of the middle ear.

In commencing the subject of diseases of the middle
ear if you will recollect that this part of the ear, closed
as it is at one end by the tympanic membrane and
opening at the other into the pharynx, is lined throughout by mucous membrane which is subject to the same
affections in this situation as in any other part of the
body (within certain limits), and that here the course
of inflammation, the changes attending it, and its products, are no different than when the same morbid
action takes place elsewhere, all the obscurity that
generally surrounds these diseases will disappear.

I feel that I am guilty of no exaggeration in saying
that to a very large proportion of students a certain
obscurity *does* surround this class of diseases, and for
the very obvious reason that students quite commonly
pay no attention whatever to diseases of the ear.

Let us first take an illustration of the simplest kind.
The occurrence of an ordinary nasal catarrh, or a sore
throat, will sometimes in its course involve the faucial
opening of one of the Eustachian tubes.   We have
seen how small the calibre of the tube soon becomes

after it leaves the pharynx. A very moderate degree of swelling of the submucous areolar tissue and an increased secretion from the surface will, you can easily understand, suffice to prevent the free passage of air from the pharynx into the tympanum. In such a case the air in the Eustachian tube and tympanum (not being replenished as usual by a constant supply passing in by the faucial opening of the tube) becomes subject to more or less absorption, the density of the air within the tube is diminished, the pressure of air from without the tympanic membrane remaining the same, the inward curvature of the membrane is increased, the chain of ossicles is rotated somewhat inwards, thus increasing the pressure of the stapes on the fenestra ovalis, and sonorous vibrations impinging on the tympanic membrane do not produce the same effect on the labyrinth as heretofore. In other words, there is deafness from obstruction of the Eustachian tube. If the balance of air is restored by suitable means, the normal hearing power instantly returns, and, supposing the inflammation to subside, the patient gets well. To take another illustration of a more severe kind. Suppose inflammatory action in the cavity of the tympanum to have proceeded to the formation of pus. If the purulent matter cannot find an exit through the Eustachian tube into the pharynx (at this time the lining membrane of the tympanum is so swelled that the opening of this tube into the tympanum is closed) it will make for itself an escape by a process of ulceration through the tympanic membrane, and there will be more or less disorganisation of

the tympanum, according to the length and severity of
the attack, before the matter has found its way out.
A series of gradations between these two extremes will
give a very fair representation of diseases of the middle
ear, each form possessing some characteristics of its
own, and varying in the effects which it leaves behind.

The term catarrh is used in speaking of affections of
the middle ear—non-purulent or purulent as the increased secretion has retained its mucous character or
become pus. More frequently than not catarrh of
either kind begins at the faucial end of the Eustachian
tube, and very often only a short part of the tube is
involved. A chronic affection of this portion of the
middle ear may remain for months without proceeding
any further, in precisely the same way that a bronchitis
will at one time be confined to the larger divisions of
the bronchi; and as sometimes in a few hours the
inflammation may spread to the capillary divisions of
the bronchial tubes, and be attended with purulent
expectoration, so may pus be secreted from the lining
membrane of the tympanum in a few hours, when (as
in scarlet fever) inflammation has spread from the
throat up the Eustachian tube. Other affections of
the middle ear commence in, and are confined throughout their course to, the cavity of the tympanum.

The illustration I have just made use of, where an
adult becomes deaf during a cold, is an example of
non-purulent catarrh being confined to the faucial
opening of the Eustachian tube, and sometimes the
subjects of this will get well in two or three days
without any treatment. The return to good hearing

is, in such a case, generally sudden and attended with a crack in the ears, the explanation of which is that as the sides of the tube, which were in apposition, become separated, the air rushes into the tympanum, and the membrane, which was before drawn in on account of the unequal pressure of the air from without, resumes its natural curvature as soon as the balance of air on either side is restored. This is the mechanical change which we attempt to bring about and endeavour to render permanent in all treatment of impaired hearing arising from such causes.

The means which we have at our command to overcome obstructions in the course of the Eustachian tube are in chief part two; viz. Politzer's method, and the catheter. The first of these two was introduced some years ago by Dr Adam Politzer, of Vienna. It consists in passing a stream of air through the inferior meatus of the nose during the act of swallowing (at this moment the opposed sides of the faucial orifice are drawn apart by the palate muscles), when the air will rush up the tubes into the tympana, and is described by Dr Politzer as follows :—" The patient, being seated, takes some water into his mouth, to be swallowed at a given signal. The surgeon, placing himself conveniently to the right of the patient, grasps with his right hand an india-rubber bag, about as large as the two fists, and introduces the nozzle of a somewhat curved, hard, india-rubber tube, movably connected with it, about half an inch into the nostril, so that its concavity is in contact with the floor of the nares. The signal to swallow is now given; both alæ are at the

FIG. 8.—POLITZER'S METHOD OF INFLATING THE TYMPANA.

same time closed air-tight over the instrument with the thumb and forefinger of the left hand, and, by a forcible pressure of the right hand, the air is driven out of the bag into the now shut nasal cavity."

This proceeding is perfectly devoid of pain for the patient; it is of course not agreeable, neither is it very disagreeable; it is beyond everything simple, and as you (this to a student), who have never before had any experience of it, try it upon this patient, a man who has frequently had occasion to have it done for him, nothing can be easier. Remember to compress the bag at the moment the patient swallows, and be particular in not permitting any air to escape from the nostril. The tube, you observe, is passed about one inch into the left nostril of the patient, the forefinger of your left hand compresses the right nostril, your thumb completing the closure of the left nostril. Having done this, the patient is quite conscious from the feeling in his ears that the inflation was a successful one. With a patient who is having this done for the first time you might perhaps have failed at the first attempt; he might not have swallowed immediately he was told to do so; just as you were compressing the bag he might have closed his mouth; you might not have completely closed the left nostril with your thumb, or perhaps, by including the tube in your thumb, have closed it. A little practice will soon overcome these very small difficulties. Although it is not, as I said, at all a painful affair, it is sometimes on the first occasion a little startling to the patient, especially if he is at all nervous; and as a perfect unison of action must exist between him and the operator, to ensure this I find it useful to let a new patient take a little water in the mouth, telling him to firmly shut it, and let him swallow at the given signal two or three times before the bag is used, taking care also to warn him of what is to be done with the bag, and at the same time assuring him that he will not be hurt. Little precautions of this kind just make the difference between a successful and an unsuccessful interview with your patient.

This applies with especial force in the case of young children. A large proportion of those who require this method of inflating the tympana are very young children, and if you frighten the child on its first visit you will never afterwards succeed in persuading it to do what you want, viz. to swallow the water when it is told and keep its mouth shut. These little patients often require a great deal of coaxing, and if you manage them well when they first see you, you will have no trouble with them afterwards. Children of four and five years of age come here constantly and have the tympana inflated in this way without making any fuss about it. .

In the place of the bag I very often with children use an india-rubber tube fitted with a mouth-piece at one end and a hard nozzle at the other, and blow through the tube as they swallow. The mode of action is of course the same, and the tube is less alarming to them than the bag. A diplomatic measure, you see, by which the desired end is attained without any sacrifice of principle.

Politzer's method is now in constant use in the out-patient's room both for purposes of diagnosis and treatment. When it is being employed, what is somewhat wordily called by some authors an " otoscope" should be used with it. A more correct expression would be a diagnostic tube. The best kind is a simple piece of india-rubber tubing, about three feet long, one end resting in the external auditory meatus of the patient, and the other in that of the surgeon. By the help of this, the air, as it rushes into the tympana, can be heard by the surgeon to impinge on the tympanic membrane. To be able to recognise this sound some little practice is necessary; and it is also advisable that a bag be used which makes as little noise as possible when it is compressed. For this reason it is much better to have the valve placed at the lower part of the bag than the upper.

If you fail to catch the sound, the question as to whether the air entered the tympana, or did not, is readily decided by the patient, who can distinctly feel the air to pass into the ears. An inspection of the membrane will also be a guide in this direction, for it will have resumed its normal curvature as the inflation took place.

Until Politzer introduced this method, the Eustachian catheter (the uses of which we shall shortly consider) was the only available instrument for overcoming Eustachian obstruction. Its place has, therefore, been to some extent supplied, and it is not necessary to use it so frequently as it was some years ago.

When an Eustachian tube is pervious, as in health, if the patient blow into the tympana with his mouth and nose firmly closed, the air can be heard through the otoscope by the surgeon to fall upon the membrane with a "thud." This is spoken of as Valsalva's method of inflating the tympana, and is useful in estimating the extent of permeability or the reverse of the Eustachian tubes. It is especially valuable where a difference exists on the two sides; and if the difference amounts to perfect closure of one with normal patency of the other, it furnishes an unmistakable diagnostic sign. It is also useful in the evidence which it affords of secretion in abnormal quantity in the cavity of the tympanum, as shown by the sounds heard through the diagnostic tube, and the appearances observable whilst inspecting the tympanic membrane as it is being done.

Some patients will succeed by this method in forcing air into the tympana on the first attempt, while others will try again and again, and fail each time. It is a good plan to explain to such patients that they are to blow through the nose when the mouth is closed, and while doing this let them hold their nose firmly between the fingers; they can hardly make any mistake about it then, and will feel the air rushing into the tympana. That at this moment the air not only enters the tympana, but that it is accompanied by an outward movement of the membrane, can be demonstrated by a very simple experiment suggested originally by Politzer. A piece of glass tubing curved in

FIG. 9.—MANOMETER.

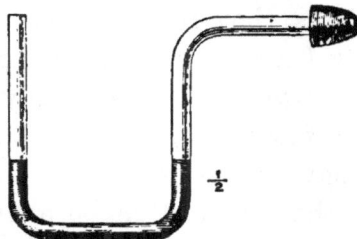

in the shape of a horseshoe, and open at both ends, has fitted to one end an india-rubber nozzle. Some

4

coloured solution is poured into the tube.   The nozzle
is fixed into the external auditory meatus, which it fits
air-tight.   As the Valsalvan method is practised the
fluid is seen to rise in the distant arm of the tube, and,
upon the patient swallowing, to fall to the original.
level.   The movement of fluid is obviously due to the
condensation of air between the tympanic membrane
and fluid in the near arm, this being brought about by
the outward movement of the membrane.   This instrument is not of much practical use, as the diagnostic
tube will demonstrate a pervious state of the Eustachian
tube or the reverse.

Obstruction of the Eustachian tube due to catarrh
of its lining membrane is a very common affection at
almost all ages, but especially with children.   As the
affection in their case is, however, in several respects
of a different character to what it is when adults are
the subject of it, we will first consider the condition as
met with in the latter class.   With grown-up persons
it is more usual to find one tube at a time affected.

The story which a man or woman who applies for
relief will tell, is that at some prior date, a few weeks
or months perhaps, they rapidly became deaf in one
ear.   By rapidly I mean, in the course of two or three
days; that soon after this, occasionally, they heard a
little better for a few hours, but that these intervals of
better hearing have never returned.   If the deafness
is very extreme there will often be a little tinnitus as
well.   You will now proceed to examine the ears, and
first look at the healthy one.   In this (I am presuming
it to be a healthy one) the tympanic membrane will

have the normal appearance already seen by you and attempted by me to be described. On looking at the other ear the difference between this tympanic membrane and the other will be very marked. The healthy one looked, as indeed it is, stretched across the opening of a cavity, but this will almost convey the idea of being stretched over objects behind it, and, in some places, in apposition to them. The outline of the malleus will be unusually distinct; very often a small artery injected with blood, and coursing from above downwards, will be visible. The short process will appear to be almost bursting through the membrane. From this point in the posterior section of the membrane will be noticed a distinct ridge separating the superior and interior division of this half into two folds. I can compare this appearance to nothing better than if a sail of a boat (the foresail for example) was supported on its leeward side by a spar, one end of which was fixed against it about its middle. The puckering produced by this object would represent in a rough manner the puckering of the membrane.

The handle of the malleus is tilted backwards. These changes in the appearances of the tympanic membrane are spoken of as an increase of the inward curvature, and I have chosen a very severe case of obstruction of the tube to illustrate the alterations that are distinctive of the affection under consideration.

You will, however, see a very large number of cases in which these alterations in curvature are not nearly so well marked, and yet in which the hearing is considerably impaired; but with a membrane that was healthy before the catarrh of the Eustachian tube set in, there are in a greater or less degree these changes to be observed.   Very slight alterations in curvature are extremely difficult to recognise, and this is because the curvature of the membrane in healthy subjects varies to a certain extent.

If you now make use of Politzer's method of inflation, and succeed in passing a stream of air into both tympana, the patient will scarcely feel anything in the healthy ear, but there will be a very perceptible sensation and noise in the affected side, and the hearing will be in a great measure (sometimes completely) suddenly restored. The change which has taken place in the middle ear will be now quite evident upon inspection of the membrane.   It will have been blown out of its proper state of tension, the folds will be gone, and it will have returned to its original plane. Before the inflation was practised, as may be supposed from the condition I have been describing, there will be undue pressure of the stapes on the fenestra ovalis; and this will sometimes induce tinnitus, which is, however, not generally of a very severe character, and by no means a constant accompaniment. When it is present, it disappears like the deafness simultaneously with the inflation of the tympana.

The patient then has experienced a sudden and great relief, and the questions arise, will this be permanent? and if not, what are the best means to make it so?

It is a most exceptional circumstance to find that the single inflation has been all that is necessary to permanently relieve the patient. I have not seen many cases where the obstruction has lasted over a week in which it has not been necessary to repeat the air douche.

With healthy adults, if Eustachian obstruction occurs in the manner related, and persists after the nasal catarrh has got well, no other treatment beyond inflation of the tympana is necessary. The frequency with which it should be practised, and the length of time necessary to continue the treatment, will of course vary according to the severity of the case. The best guide in the matter will be the length of periods that the improvement in hearing remains after each occasion. This will sometimes only last a few hours (perhaps the morning following the inflation the patient will be as deaf as ever), and the case will then give some trouble. The good hearing may, however, remain for two or three days; and these patients will only require the air douche a few times.

This at least is the general progress of such cases, but of course no rule can be laid down, and every now and then a patient will be under treatment for many weeks.

The portion of the tube first affected was, as we saw, the faucial opening, and in some instances the whole throat, including this part, will remain for a long time in the condition we are accustomed to speak of as relaxed. In such instances very great benefit often follows the use of astringent local applications to the throat, and in applying them the object should be to touch freely the orifice of the Eustachian tube. To accomplish this it is convenient to use a brush with a handle curved at right angles, the same kind indeed as is used when the larynx is to be touched. Chloride of zinc ₃ss to ₃j, or the salt perchloride of iron, ₃j to ₃j of water, are what I generally recommend. I dare say any powerful astringent would do equally well, but these are what we use here. In this class of cases it is almost needless to say that the throat should be inspected, and, if possible, the orifice of the Eustachian tube be examined with a laryngoscopic mirror. To gain facility in the examination of the posterior nares requires considerable practice, and it is certainly not so easy to see this part as it is to see the larynx. If the pharynx is of a good depth, measuring from before backwards, the orifice of the Eustachian tube is far more readily seen than if it is less capacious. If the pharynx is very short from before backwards, it is extremely difficult to see the opening of the Eustachian tube, at least I find it so much so that in such cases I very often cannot manage to get a view of it at all. An enlarged tonsil will be in the way for this kind of examination. I advise those of you who wish to learn how to use this instrument to begin upon a trained patient; by this I mean one who has very often been examined before, for his pharynx will tolerate manipulation far better than a new case will be able to do.

The base of the tongue should be well depressed by the forefinger of the left hand. The mirror should face upwards and be placed between the tonsil and uvula, as low down as possible, and nearly, but not quite, touching the back of the pharynx. If it actually touch the pharynx the patient will retch. Let the mirror now be turned the least bit outwards, and the reflection of the orifice of the Eustachian tube will be seen at the top of the mirror. This opening is triangular, the edges are well defined, and even in health there is very often a piece of mucus just at the orifice. If this part is relaxed, the lips of the opening look raised and pouting.

A most useful adjunct to the occasional or regular use of Politzer's method of inflation will be found in the application of alkaline solutions through the nostrils. The most convenient method for patients to adopt is to simply make use of any form of ordinary syringe. The nozzle of the syringe should be placed in the inferior meatus of the nose, and the fluid be very gently injected through to the posterior nares ; some of this will pass out again through to the opposite nostril, while some will enter the mouth, and should be spat out. This is a better plan than one sometimes followed, viz. the use of the nasal douche, and for the very good reason that this is not altogether free from risk. In its application some of the fluid is apt to pass up into one of the tympana, and then inflammation may be excited in this cavity and the tympanic membrane may give way as a consequence. I have not infrequently been consulted by patients with whom this accident has occurred.

Before concluding this lecture I have four cases to show you, and before examining them I may say that in estimating the variations in hearing power, the ticking of a watch, as being convenient and always at hand, is generally the sound employed. The facility with which this heard, however, gives no proximate measure of the extent of deafness.

Persons will be found who, though unable to hear a loudly ticking watch more than a few inches from the ear, are not perceptibly deaf to conversation; and again, a person whose hearing is very considerably impaired for general sounds will sometimes be able to hear a watch at many feet from the ear. It is therefore only useful in estimating changes in degrees of hearing in a patient under treatment who is seen from time to time. Even then a more correct guide is the patient's own estimation of voices and sounds which he is in the habit of hearing. With young children it is not easy to find out whether the hearing improves, or the reverse, and the most reliable proofs of either is the manner in which they reply to questions which are put to them when their heads are averted, and they are not therefore expecting the trial. Both with children and adults it is a good plan to test the hearing by making them repeat after you letters, numbers, or words, and in doing this they must not be allowed to watch the lips of the speaker, as many persons gain great assistance by reading from the lips, and so might be considered to hear better than they really did.

No. 1. J. E——, a married woman, æt. thirty, came here on the 14th March last. The hearing of the left ear had been lost almost completely during childhood, and exactly one month before she applied she became deaf without any cause which she can specify in the

right ear.   In the course of two or three days she was
as deaf as on the day she came here.   I mention the
fact of the greatly impaired hearing on the left side,
as if she had heard well with that ear she would not
have applied for relief till much later, and as a rule so
long as people hear with one ear there is no unpleasant
symptom they will suffer so long without seeking help
as deafness; indeed, this is one of the chief reasons
why in such a considerable portion of the patients who
come here, the diseases of the middle ear are of so
long standing, and oftentimes so intractable.   The
right tympanic membrane had the puckered appearance I have just described; she required a raised
voice at a few feet from the ear, and the watch was
heard one inch from the auricle; air passed freely into
the tympanum on Politzer's plan, increasing the
hearing to eighteen inches with a watch, and conversation was heard as well as before the attack.   The
next week the hearing was not nearly so good as it
had been after the inflation, and the watch was heard
at five inches.   Another inflation again improved the
hearing as before.   The same treatment was pursued
on the next Thursday, and on the next.   This time,
her fourth visit, the hearing has been maintained as it
was on the previous week immediately after the inflation.   The following week the hearing was the same,
and it has not since then changed.   To look at the
tympanic membrane to-day no one could say that the
middle ear had in any way suffered.

The next case, No. 2, is also a woman, æt. thirty,
who is at present under treatment.   Three weeks ago

she came here saying that a week before she could hear well, and that then she became deaf in both ears. She was extremely deaf, and the watch was heard at one inch from the right ear, and not with the left except upon close contact.

There was no history of any sort. Both tympanic membranes had similar appearances to what was seen in the last case, and at this moment such appearances are present though in a less degree.* Now that the tympana have been inflated, they have resumed their proper plane, and she hears very well. From the gradual improvement she has made, the catarrh is evidently subsiding and she will soon be well.

Case No. 3 is in many respects different from the other two. G. C—, æt. thirty-four, had good hearing in the last week in May, and came here to-day, June 13th, very deaf with both ears, a syphilitic sore throat, and an eruption of the same nature on his chest, arms, and face. Both membranes have a somewhat increased concavity ; he can only hear a watch pressed against either ear, and on inflating the tympana, the hearing distance for the watch is increased to six inches with the right ear and three inches with the left. The impaired hearing is not wholly due to obstruction of the Eustachian tubes ; it is so in some degree, as shown by the change for the better produced by inflating the tympana ; but a considerable amount of the deafness depends upon the constitutional syphilis. We shall discuss by-and-by the syphilitic affections of the ear, but at present it will be enough to say that he will in

* Politzer's method was then used.

all probability recover under medicines from the extent of deafness which depends upon the syphilis so far as it affects the nervous structure of the ear. We are now concerned about the local change from health in the Eustachian tube, and for this local origin of the impaired hearing, the treatment he will be subjected to will consist of inflation of the tympana at regular intervals and the application of chloride of zinc solution to the pharynx. When the air douche has no further immediate effect upon the hearing, whatever amount of deafness remains may be considered due to the constitutional syphilis, the proper treatment of which calls for no remarks from me.

The next case is that of a boy, J. H—, æt. 19, who became deaf during a cold with both ears, and remained so a month after he had quite recovered from the cold. The first inflation increased the hearing power from six inches with the watch to three feet, and the improved hearing was maintained for six days without a repetition of the air douche. This was quite sufficient to make it highly probable that he would not be long under treatment. The tympana were inflated on two other occasions at intervals of a week, and without any further treatment the catarrhal condition subsided and he heard well.

LECTURE III

FAR more common than with adults is the Eustachian obstruction met with in children, especially those of a strumous type. In these subjects a vast extent of mucous membrane is in the same condition—the nares, fauces, and Eustachian tube being affected together. The mucous membrane throughout is thickened and tumid, and secretion from the surface is much more abundant than it should be. The tonsils are generally enlarged, and sometimes to a very considerable extent. Of such children it is hardly necessary to ask what is the matter. The stupid vacant look as they advance with open mouth, and their generally flabby appearance, proclaim their disease. They snore loudly in their sleep, and the deafness is generally severe. The tympanic membrane will be seen to be drawn in, but to retain its proper translucency; as a rule there is no tinnitus. It is surprising for what a length of time this state of things will go on with children and yet permit of complete recovery; while, on the other hand, in the case of adults, when the Eustachian tube is obstructed from a relaxed condition of the mucous membrane, the tympanum will generally become involved if they are not attended to soon after the deafness is noticed. The treatment to be adopted for these young patients is to apply astringent solutions locally to the fauces, tonic medicines (preparations of iron and the mineral acids), plenty of fresh air and exercise, together with the ordinary rational means of improving the health. The tonsils, however much they may be enlarged, cannot operate mechanically by closing the Eustachian tube, but they may interfere with respiration, and, by presenting a large surface of thickened mucous membrane, assist in keeping up the unhealthy condition. These considerations should weigh in deciding whether they should be removed or not. Of all things it is important that the Eustachian tube should be regularly opened; if this is not done, all else will be useless. To accomplish this, Politzer's is the best method. Measured by the good it has done, it would be difficult to overvalue this proceeding in treating diseases of the middle ear; for before this plan was suggested, a large number of cases, especially children and timid women, on whom catheterisation was not practicable, went unrelieved.

After the inflation, although the hearing is at once very much improved, the deafness will not unfrequently return in a few hours. It should, however, be assiduously persevered with; and as the mucous membrane gets into a more healthy condition, the periods of good hearing will become longer and longer, and improvements which at first lasted only for six hours will after a few times remain for twenty-four, and so on. When the Eustachian tube admits of air passing through it in the natural way, the ossicles and tympanic membrane will remain in their proper position. No trouble on

the part of the patient and surgeon should be grudged
to ensure this, as good results will assuredly repay
labour thus bestowed.   With some of these children it
may suffice to inflate the tympana once or twice a
week, while with others daily inflation for a few times
may be necessary; but, sooner or later, the hearing
will be completely restored.

Cases of this kind are of extremely common occurrence, and I take for illustration two which will give
you an idea of their general course.   L. F—, a boy,
æt. 5, was said by his mother to have been deaf, more
or less, for three years before I saw him on 17th April.
He was very deaf for conversation, and his speech was
beginning to fail, as it always does with very young
children who are deaf to any considerable extent.   The
watch was heard about four inches from either ear.
Both tympanic membranes had the normal translucency; but the inward curvature was exaggerated.
Politzer's inflation increased the hearing to three feet
on either side.   The throat was relaxed; the uvula
pendulous; he breathed almost entirely through the
mouth, and snored in his sleep.

The next day the good hearing had very nearly been
maintained.   Politzer's inflation was used every other
day for three weeks; twice a week for two weeks, and
once a week for a month.   On each occasion the throat
was touched freely with solution of perchloride of iron.
Besides this he was taught to syringe through the
nose a weak solution of soda.   This was done in order
to cleanse the lining membrane of the nares, so as to
make it more easy for him to breathe through the nose,

and was thus a safe substitute for the nasal douche. I referred to this in the last lecture. It is very useful indeed when the mucous membrane of the nose is thickened and habitually secreting more than natural. At the end of nine weeks he was dismissed with good hearing, and on October the 9th his mother told me that he had had no further treatment and continued to hear well.

The next case was more tedious, but in its termination equally satisfactory. I give it in detail, as it is most instructive in showing how long this tumid condition of the mucous membrane will continue, how extreme the amount of deafness, and how necessary it is when the affection is clearly recognised to enforce perseverance with the patients and their friends.

September 7th, 1870.—F. P—, æt. 15, a delicate-looking boy, with pale flabby cheeks, had lived in India as a child, and four years before had come to England to go to school. The account which his friends gave was that up to three years ago he had heard quite well; from that time he had been getting deaf. There was no distinct history of cold as an origin, but he was worse when he had a cold. The fauces were red, the tonsils somewhat enlarged, and the uvula hanging down and lax; the lining membrane of the nose red and swollen. He breathed chiefly through the mouth, which he never kept shut, and snored loudly at night. Required a raised voice near him to make him hear. The watch was not heard excepting in contact with each ear. Both sides, as far as hearing was concerned, were alike, as also was the appearance of the tympanic

membranes. They were of the natural colour and translucency, and the concavity very considerably increased.
He had been getting much more deaf for the last six
months, and his education was of course very much
impeded. He with difficulty forced air into the tympana. With Politzer's inflation air entered with a
bang, the hearing immediately rising to three feet with
the watch on each side. This improvement lasted during
the whole of that day till he went to bed, and on the
next morning he was, he said, as deaf as ever. His
nose was always clogged with mucus. He was ordered
every morning to draw up the nose half a tumbler of
water containing half a teaspoon of table salt into the
throat, and to spit it out. To use a gargle of alum
ten grains to one ounce twice a day.

In the first week Politzer's inflation was practised
daily, and the intervals were gradually lessened, so
that by November 26th, eleven weeks after he was
first seen, the good hearing was sustained for several
successive days, and he could at this time hear conversation across a large room without the voice being
raised. He was more or less under treatment for four
months, and by the following spring he could hear
well, and required no further treatment.

To my mind the case most strikingly illustrates the
great advances which have been made in treating disease of the middle ear of late years, and which may be
said to be entirely due to the German aural surgeons.
Some years ago the attendance of the patient for inflation by the Eustachian catheter would have seemed
both to him and the surgeon interminable; in all pro-

bability the one or the other would have given it up
in disgust before much benefit had resulted, and the
hearing would have relapsed into its former condition.
Do not suppose from this that we have done with the
Eustachian catheter; on the con-
trary, its use is more than ever
found to be necessary in appropriate cases, but as a rule the use
of the Eustachian catheter is required more frequently in affections
of the tympanum than in those of
the Eustachian tube; but sometimes,
when the obstruction of the tube is
so severe as not to yield to other
methods, it is necessary to employ
it; and, in addition to inflation of
the tympanum by means of it injections are sent into this cavity. The
catheter may be made of either
silver or vulcanite. The first kind,
A, can be bent to any curve required, and the latter, B, are inflexible except on being exposed to
heat, when they can be curved as
may be necessary, and become
inflexible on cooling. So long as
the catheter does not bend while
being used, it is not important
what it is made of. Four or five different sizes are
quite enough. The metal ring shows the position of
the instrument in passing it—whether the curve is

FIG. 10.

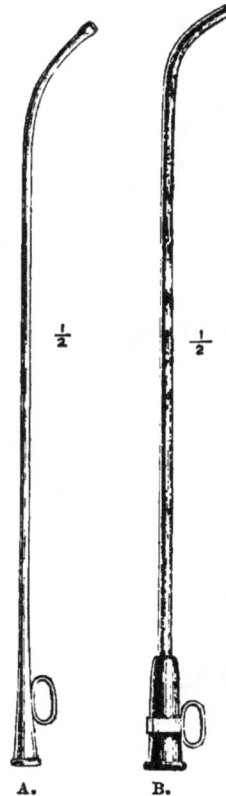

turned to the right or left, upwards or downwards. The other end is made somewhat conical, so as to fit the nozzle of an india-rubber bag which is employed in injecting air or fluid into the middle ear.

The mode of passing the catheter is as follows :— Place the patient in a chair, and let him lean back in it; steady his head with the left hand firmly fixed on the top of it; hold the catheter lightly in the right hand, with the curve downwards, and pass it quickly in this position through the inferior meatus of the nose to the posterior wall of the pharynx.   When this is felt, withdraw the catheter about half an inch, and tilt the point of the curved end rather upwards and to the left or right according to the side which is being operated upon.   Now hold the catheter and end of the patient's nose steadily between the thumb and the first two fingers of the left hand.   All this time the ear of the patient and surgeon are connected by the diagnostic tube.   The point of the catheter is now supposed to be in the pharyngeal orifice of the Eustachian tube, but observe please, that the only certain sign of this being the case is that when air is forced into the catheter it will be heard through the otoscope to impinge upon the tympanic membrane when a stream of air is passed down the catheter.   (I am presuming that the Eustachian tube is not occluded.)   This is the most common method of using the catheter, and I think it is the best.

A plan was suggested some little time ago by Dr Löwenberg.   The catheter was passed, as before, to the back of the pharynx; the point was then turned

inwards, and withdrawn till it stopped by being
hooked round the vomer; the catheter was then
turned completely round, and so the position of the
orifice of the tube determined. To learn to pass an
Eustachian catheter, it should be done first on a
vertical section of the head (a moist preparation) in
the way first described, so that each movement can be
watched, and the opening of the Eustachian tube seen.
A cloth may then be arranged so as to hide the opening
from view, and removed so see if the point of the
catheter is in the right place. After this it should be
practised on the living. The chief mistakes made by
beginners are, not keeping the catheter close to the
floor of the inferior meatus of the nose, thus getting it
into the middle meatus, and, when it has reached the
pharynx, not withdrawing it sufficiently before it is
turned outwards. It is well to get into the way of
passing it through the meatus as rapidly as possible,
for this is to the patient the most disagreeable part of
the operation. It is remarkable how little discomfort
it causes to some persons, and how much others object
to it. Although a somewhat disagreeable, it is not a
painful proceeding, and after the catheter has been
used a few times it is tolerated without much inconvenience.

A stream of air may be sent through the catheter
either by blowing through it with the mouth or using
the india-rubber bag.

The advantage derived by using the breath, on account of its temperature, is not appreciable; so I never
use the breath. The bag is more convenient; a

stronger stream of air can be sent by its means, and if injection of fluid is required, its employment becomes necessary. Occasionally some considerable difficulty is met with in introducing the catheter. This is chiefly the case when the septum of the nose is placed obliquely, and the deviation is generally towards the left side. If this be so much as to prevent altogether the passage of the catheter, and if it be imperative to use it for this side, the catheter must be bent to a greater curve, and introduced through the right inferior meatus of the nose. It is troublesome to manage it in this way, but it can be done, and is the only practicable method in such a case. If the inferior turbinated bone is much larger than usual, this will also make the meatus impassable for the catheter.

Before using a catheter it is well to make a rule of blowing through it, so that you may be sure that it is not stopped up. There is no need, however, to smear it with oil, as the surface over which it has to pass is well lubricated with its natural secretion. One other caution, that is perhaps a needless one,—be careful that the catheter is always thoroughly washed after it has been used, for there have been instances in which patients have become syphilised through the neglect of this precaution.

The chief province of the Eustachian catheter is in the treatment of those affections of the middle ear for which Politzer's method is not adapted. Such occasions are of daily occurrence ; for instance, where one tym-panum only is affected and it is not advisable to in-flate the other ; where a prolonged stream of air is desirable in place of the sudden gust which results from Politzer's method, and for many other conditions which we shall consider as we go on, not only where the air douche is necessary, but as a medium through which further local treatment can be applied to the middle ear. How far its employment is necessary for purposes of diagnosis is a question which will be answered in a different manner by some to what it will be by others. Although in using it we often gain additional information as to the condition of the tympana, I must confess to thinking that in a general way it is seldom necessary for a correct diagnosis. It is a rare occasion where we are not able to discover by other means that have been mentioned (Valsalva's method, or Politzer's) obstruction or occlusion of the Eustachian tube, and variations in the condition of the tympana from extreme dryness to collections of fluid in this cavity; after all it is the relative freedom or difficulty with which air enters the tympana, and proportionate degrees of and entire absence of secretion that are the chief points to be decided by auscultation of the middle ear. At the same time, since in most cases of chronic disease of the tympana the catheter is very frequently used in the course of treatment, as the case progresses changes in the sounds produced on inflation are oftentimes observable, and deserve careful attention; but I am quite unable to understand, or rather I should say appreciate, such an opinion as the following given by Dr Kramer in speaking of this subject. " The physical characters of these acoustic symptoms are far more clear and distinct than those which present themselves when the stethoscope is used for the diagnosis of the diseases of the respiratory organs."

The permanent effects to be apprehended from obstruction of the Eustachian tube, either · at the faucial or tympanic orifice, are always due to neglect. Persons get deaf during a cold, remain so for a few days, and recover without treatment. In future, they or those who have witnessed this state of things very naturally argue that this is the usual course of events. It is not so. Any obstruction that has persisted for many days (this applies chiefly to adults) is liable to remain more or less, or at any rate to make necessary a long course of treatment which would not have been required if attention had in the first instance been given to the case. In other words, a few inflations of the tympanum at the commencement would have placed the patient in the way of recovery. It too often happens that weeks or months are allowed to pass by before relief is applied for, and it is then found that air passes by Politzer's method very imperfectly into the tympanum. Regular inflation should then be practised with the Eustachian catheter. After a little time air will pass more readily, and the hearing improve proportionately. In some instances at this clinique, after the catheter has been used for several days successively with only slight relief, almost perfect hearing has followed some single inflation. This has been noticed especially when a few drops of a warm solution of soda have been injected into the tympanum the day previously. It appears probable that, the obstruction being at the tympanic orifice of the Eustachian tube, there may have been a small plug of mucus, which, after having been softened by the injection, has been dislodged by the next inflation. It is in some degree confirmatory of this hypothesis that in these cases the hearing remains good without any further treatment. This is a highly mechanical explanation of the phenomenon, but it must be remembered that in these cases impaired hearing depends on a mechanical lesion, and that when we are dealing with what we cannot see some speculation as to cause and effect must be permitted.

Non-purulent catarrh of the tympanum may commence in the manner described when speaking of Eustachian obstruction in adults dependent on a relaxed state of the mucous membrane of the throat, and thus the cavity of the tympanum may become involved by extension of the catarrh in this direction, or the tympanum may become affected independently of the Eustachian tube. In either case, repeated attacks of this kind, if allowed to proceed unchecked, are among the commonest causes of confirmed deafness; this symptom being due principally to accumulations of mucus within the tympanum (fluid in the early stages, and becoming inspissated in the later), which by their presence interfere with the proper performance of the functions of the ossicles and tympanic membrane, and also with the conduction of sound through the tympanum to the labyrinth. The course of the affection is somewhat as follows (taken from the case-book):—A healthy man, æt. 40, could always hear well until six days ago; he then became slightly deaf on the left side,

with some pain in the ear and a feeling of irritation in
the meatus. In two or three days the pain, which was
never severe, subsided. When seen, he was very deaf
to conversation, and could hear a loudly ticking watch
at one inch from the ear. The tympanic membrane
was opaque, and drawn in more than natural; the
short process of the malleus was rather prominent, and
a few injected vessels could be seen near the handle of
the malleus. The meatus was red and rather tender
on introducing the speculum. Gentle inflation was
practised every day, on Politzer's method, the hearing
improving after each occasion, and in a week he had
quite recovered. At first the air as it entered the
tympanum did so on the affected side with rather a
moist sound, but when he was dismissed it passed into
both tympana alike. In this case, the swollen state of
the lining membrane had prevented the free ingress of
air into the tympanum, and by keeping this cavity inflated until the inflammation had subsided, any permanent injury to the hearing was prevented.

There are a few points which are noticeable as presenting a decided difference to what was seen when the
Eustachian tube was affected to about the same extent
as the tympanum was in this case : the absence of any
throat affection, the occasional and slight pain in the
ear, the evidence of secretion in the tympanum as heard
by the otoscope, the variation from translucency in the
membrane, the injected state of its vessels, the redness
and tenderness of the meatus. The inflammation
beginning in the lining membrane of the tympanum is
thus seen to involve in turn the three layers of the tym-

panic membrane and the meatus. This is the general course of events,—all affections of the tympanic membrane, with very rare exceptions, being secondary to morbid action which has begun either in the tympanum or the meatus. Another appearance seen in these cases is a lessening of the area of the tympanic membrane, which is due to the swelling of the external meatus immediately around, encroaching on the membrane. If the symptoms in the case just related had increased in severity, and the secretion been more copious, the pressure caused in this manner might have resulted in a rupture of the tympanic membrane, even if the secretion had still been strictly of a non-purulent character. One stage farther in the course of inflammation, and the secretion would have been purulent. Although sometimes the non-purulent merges into the purulent variety of catarrh of the tympanum, they are commonly described as separate diseases. Purulent catarrh generally accompanies one of the exanthemata, or it comes on without any apparent cause in a very sudden manner, and runs a very rapid course, the membrane giving way in a few hours.

The case which I have just taken by way of illustration was of an extremely mild character, very brief in its duration, and a complete recovery took place after treatment of the most simple kind. This rapid and complete recovery is very exceptionable, and for the very good reason that the patients are very seldom seen at so early a stage of the affection. It is one thing to have to deal with a tympanum which has been the seat of catarrh for a few days, in which the mucus secreted is slight in quantity and recently effused, and a tympanum that has only for a short time been deprived of the air with which in health it is filled; but a very different matter is it to have to treat a case where the supply of air in the tympanum has not been replenished for many weeks, and half filled perhaps with viscid mucus. Although, too, there was slight pain in the ear in this man's case, this symptom is not always present to warn the patient to seek advice; it is so, however, whenever the tympanic membrane shares the condition of the rest of the tympanum, and this would in the course of things be expected, for, as you know, it is a dense fibrous structure plentifully supplied with nerves.

Having premised so much, and given a sketch of Eustachian obstruction, and of an attack of non-purulent tympanic catarrh, commencing either in the tympanum or proceeding from the throat, what are the effects which have been observed to remain after death in persons who have been the subject of non-purulent catarrh of the middle ear? Speaking generally, these evidences of disease are more or less obstruction of the Eustachian tube, in some cases occlusion; thickening of the lining membrane of the tympanum; moist or dry mucus, either as collections around the ossicles, bands between the ossicles themselves, or between the ossicles and sides of the tympanum; anchylosis of the ossicles, either between each other or between the stapes and fenestra ovalis. Patients who apply for relief after successive attacks of non-purulent catarrh give the same history with some variety of details;

periods of deafness occurring at intervals when suffering from cold, or always worse hearing at such
times, sometimes accompanied with slight pain in the
affected ear; occasional tinnitus. Not too much
dependence must be placed on the latter symptom, as
it will be shown in another lecture to be present under
so many circumstances; but, as a rule, with these
cases it is not a favorable symptom. The tympanic
membrane will present certain alterations in curvature
and translucency. It requires long practise to recognise changes in curvature. In placing side by side
two patients, in one of whom the membrane is healthy
and translucent, whilst in the other it is opaque, the
first impression will be that the healthy is the more
concave of the two, when in truth, perhaps, it is the
reverse, the change in colour giving it a more flattened appearance. The curvature, too, varies somewhat in individuals, as also does the inclination of the
membrane. Of all appearances indicating an abnormal
concavity, the most easily recognised is an unusual
prominence of the short process of the malleus. Opa-
cities in the tympanic menbrane depend upon thickening, and in proportion to the seat of this so will
opacity be valuable in diagnosis. As stated in a
previous lecture, it is sufficient to regard the tympanic
membrane as consisting of a fibrous layer, on either
side of which is an external or dermoid layer continuous
with the lining of the meatus, and a mucous layer
continuous with the lining of the tympanum. In
catarrh it is the last-named layer that is the seat of
thickening; but as either of the other two layers may

be the one affected, an opacity, whether it be a general one, or only involve a portion of the membrane, will not settle a question in diagnosis, but when taken in conjunction with other symptoms, and the history of the case, it is valuable as confirmatory evidence in arriving at an opinion in a case. A patient may have a thoroughly opaque tympanic membrane, produced, perhaps, many years before we see him, by a catarrhal condition, which was completely recovered from, leaving him with a good hearing, and none the worse, except this change from translucency and lustre in the membrane of the tympanum, the impaired hearing, for which subsequently he applies for relief, depending upon a nervous lesion, entirely unconnected with the tympanum.

I know that I shall be challenging criticism by saying that it is a point quite open to question whether thickening of the membrane is of itself ever a cause of impaired hearing. Physiologically speaking, a man with an abnormal increase in the substance of one tympanic membrane ought to have imperfect hearing as a consequence of such thickening; but what am I to say when I find that this structure is subject to the most extensive increase of its substance by the deposit of calcareous matter, and that this condition has been observed in many cases where the hearing has been quite good. Undoubtedly in most instances of such deposits (they consist of phosphate of lime) there has been a history of tympanic disease, and the patients are more or less deaf, some extremely so; but it appears probable that the loss of hearing is due to

changes within the tympanum, and not in the membrane. At any rate it seems fair to infer this when so
many persons have these deposits in the tympanic
membrane and hear well. This conclusion has been
further strengthened where opportunities have been
afforded of examination after death, of the ears of
persons who have had such changes in the tympanum;
and whose power of hearing has been noted. In these
cases where there has been deafness, either slight or
extreme, it has been amply accounted for by morbid
conditions situated behind the membrane. This is
particularly noticeable in a collection of pathological
specimens in the General Hospital of Vienna, belonging
to Dr Politzer, and which I had an opportunity of
examining in 1871.

LECTURE IV

IF I were asked, do patients ever recover altogether
from the effects of chronic catarrh of the middle ear?
and, what do you think the best treatment in such
cases? I think I should reply somewhat as follows :—
To the first question, if patients are seen during the
first attack of uncomplicated non-purulent catarrh of
the middle ear, however severe the attack may be,
provided the tympanic membrane has not given way ;
however great the impairment of hearing, with ordinary
patience and care in conducting the treatment, it is a
very exceptional circumstance for them not to recover
the hearing, either completely, or nearly so. For the
rest, those in whom recovery is only partial, or does
not take place at all, are they who either have not
been attended to during the first attack, or, perhaps,
have had several in succession without being submitted
to treatment.

To the second question I would reply that all treatment should have the same objects in view. By inflation of the tympanum with Politzer's bag or the
catheter, to overcome any obstruction to the free passage of air into the tympanum, and to keep this cavity
supplied with air until the swelling of and extra
secretion from the lining membrane (for it is this

condition which causes the obstruction) shall have
subsided ; by the injection of fluids into the tympanum
to improve the condition of its lining membrane ;
under appropriate circumstances by a timely incision
in the tympanic membrane to prevent its rupture being
caused by pressure from within, and taking advantage
of the opening thus formed to bring about the expulsion of mucus which may be filling up the tympanum.

I do not think I should be far wrong in saying of
tympanic affections that there are very few diseases
in which the results of treatment have been more unsatisfactory up to a comparatively recent period, and
this is in great measure to be accounted for by the
fact that their pathology was imperfectly understood.
The careful and laborious dissections of Mr. Toynbee,
by demonstrating the morbid appearances met with in
the cavity of the tympana have been of great service,
and I cannot but think that in a very considerable
degree they have been the means of suggesting to other
surgeons rational measures for relieving such conditions, and here I may say that we are chiefly indebted to the following surgeons for reform in the
treatment of tympanic disease,—Drs Von Tröeltsch,
Adam Politzer, and Gruber in Germany, and Mr
Hinton in England.

In the case of tympanic catarrh to which I directed
your attention at the last lecture, the deafness was not
very extreme, and the patient began to improve so soon
as he was placed under conditions favorable to recovery,
and he was well in the course of a week.

It is unusual to have so little trouble with these

cases, and very often the patients are under treatment for several weeks. If the Politzer's inflation restores the hearing but imperfectly at the time and does not give a very thorough inflation to the tympanum, the air douche should be given with the catheter, and in doing this the bag should be compressed at least five or six times successively. The completeness of this measure will show itself by its immediate effect upon the hearing, and I can give you no rules which will universally apply as to how often it will be necessary to use the catheter, except by saying that if the improved hearing is very transient, it will require to be repeated more frequently than if it lasts for several days. Experience will guide you in this respect, and after treating a few cases you will be able to judge pretty well how often the tympanum ought to be inflated in this way when a few days have elapsed after the first occasion on which the catheter has been used.

Some cases will gradually get well if nothing more than this treatment is pursued every third or fourth day for a few weeks ; with others more vigorous measures will be necessary, and here you must be guided chiefly by evidence which you can obtain as to the amount of secretion in the tympanum. If upon inflating the ear either by Valsalva's method or the catheter a distinct mucous râle in the tympanum can be detected, very considerable benefit will follow the employment of a weak astringent injection into the tympanum.

The usual method of injecting solutions into the cavity of the tympanum is by means of the Eustachian catheter. This is passed in the ordinary way, and steadied with the left hand ; about half a drachm of the solution is dropped into the open end of the catheter with a glass syringe ; the india-rubber bag (which has been spoken of) is immediately fixed into the catheter and compressed. Although the level at which the Eustachian tube enters the tympanum is rather near the roof, it is not found that after the fluid has entered the tympanum in this way enough collects at the bottom of this cavity to do any harm. It is not as if the fluid entered the tympanum in a stream. It leaves the catheter in a dispersed jet ; some of it then is lost in its passage up the Eustachian tube, and by the time it has entered the tympanum, and moistened its walls and contents, there will not be much to sink down to the floor of the cavity. It is more useful to notice the results of this treatment than to theorise on what takes place. The fluid can be heard through the otoscope to enter the tympanum, and the patient is quite conscious of it so doing. Another plan was suggested by Dr J. Gruber, of Vienna. The head of the patient is inclined to the side to be operated upon. A little of the fluid is passed with a small syringe into the inferior meatus of the nose ; the patient then, with the mouth closed, and compressing the nostrils between the forefinger and thumb, forcibly inflates the middle ear, with the effect of sending some of the fluid into the tympanum. Intelligent patients can do this ; but it has the disadvantage that they are not able to regulate the amount injected, and for this reason I think it not quite a safe proceeding. Too much fluid forcibly injected in this way is liable to excite inflammation of the tympanum.   A third method in constant use at this hospital is a modification of the last one.   The fluid is in the same way passed into the inferior meatus of the nose, and the patient holding some water in his mouth, the tympana are inflated on Politzer's plan, the nozzle of the bag being placed in the meatus which contains the fluid, and the head bent as before.   This is found peculiarly applicable in the case of children where either of the two other methods is attended with considerable difficulty.   The cases adapted for astringent solutions are those in which there is an excess of secretion, as shown by the moist sounds conveyed through the otoscope on inflating the tympana.   A very useful solution is one of sulphate of zinc, about two grains to the ounce of warm water.   Anything much stronger than this gives rise to a good deal of pain.   The injection should be repeated every three or four days, and continued for some time.

The case to which I direct your attention to-day is a very good example of the course which these sort of cases take, and the general routine of treatment that I find most serviceable for them.   The affection was more than usually intractable; it was not the first time the patient had had trouble with the ear, and he had delayed for three months before he sought advice.

C. H—, a healthy man, æt. 30, gave the following account of himself.   Twice during the previous two years he had been deaf for three or four days at a time with the right ear, but had recovered without any treatment.   In the past three months he had been deaf on the same side, and suffered from a constant noise in the ear. The hearing on the left side was good. With the affected side the watch could be heard at three inches from the ear, but with the healthy ear closed he required a voice considerably raised, and within a few feet from the ear; in short, he was extremely deaf on that side. The tympanic membrane was very much collapsed, and had an appearance as if fallen in on to the promontory; the posterior section was in ridges, the short process of the malleus very prominent, and the handle not traceable. The lustre of the membrane was gone, and altogether it had such a distorted look that any one, even if he were not much in the habit of examining tympanic membranes, would have observed on comparing the appearances upon the two sides that there was something very wrong with the right side.

When a vibrating tuning-fork was placed on the top of the head the sound from this was very much intensified on the deaf side. Without staying now to explain all the uses of the tuning-fork for purposes of diagnosis, I may say that vibrations of sound thus conveyed produce their effect independently of the conducting apparatus upon the auditory nerve, and their effect becomes intensified by closing the meatus, or, as in the case before us, by the presence of any obstruction in the tympanum. A little air was passed into the tympanum by Politzer's method and increased the hearing slightly, but not very much. On using the catheter, and thus sending a full and continuous stream of air into the tympanum, he experienced a sense of great relief, not only in respect of the hearing, but the noise in the ear ceased. The hearing distance of the watch was thus increased to nearly three feet, and the voice was heard very much better.

Three days later he reported that he had heard well for the two days after he was first seen, but that now he was as deaf almost as before. The catheter was used for two weeks every third day; sometimes the good hearing would last for four days, at one time it did for five successive days, at others he would become again deaf the day after the inflation, but the tinnitus did not return (neither did it all through the treatment) after the first time the air douche had been used. The sound which could be heard while the air was blown into the catheter was very distinctive of mucus in considerable quantity in the tympanum. At the first few compressions of the india-rubber bag it was almost a gurgling sound, and the man himself felt as if there was something in the ear. It seemed as if the stream of air dispersed the secretion, for the sound was less moist (so to speak) after four or five separate jets of air had been sent through the catheter. He did not appear to be making much real progress, and so I now injected a little warm solution of sulphate of zinc (three grains to an ounce of water) through the catheter every other day for two weeks. All this time he steadily improved. Thirteen days afterwards he was again seen and had had no relapse. His hearing on the right side was distinctly not so good as on the other, either for a watch or for a very low voice at any distance, but for practical purposes, such as ordinary conversation or hearing a speech in a public place, it

was quite good enough, and he was not conscious of
being at all deaf. With this man there was no history
of colds and sore throat, so it is very difficult to say if
the affection began in the throat or in the tympanum ;
at any rate, it was the tympanum that was at fault
when he came to me. At the time he was dismissed
the membrane showed signs of the tympanic affection.
The folds in the posterior section which were so marked
when he first came had disappeared, and the membrane
generally had returned from the collapsed condition to
its proper plane, but it was dull white in part, and
discoloured, evidently very much thickened, and the
outline of the handle of the malleus was not distinguishable.

In the course of catarrh of the middle ear, when the
secretion within the tympanum is very copious, the
pressure which this produces will induce thinning of
the membrane. In most cases this thinning will be
general, but in others it will be circumscribed. In
either there is a certain amount of risk lest the membrane give way, and it is partly to guard against this
mishap, and partly to form an outlet for the mucus
when it will not disappear under other plans of treatment, that an incision is sometimes made into the
membrane, and the fluid evacuated through the cut
upon passing air into the tympanum (of this treatment
I shall presently speak). Another symptom or effect
of fluid in the cavity of the tympanum is that patients'
hearing will vary much with changes of position.
Thus, while lying down, they will be able sometimes
to hear fairly well, and on rising will in a few moments

be as deaf as before. The head bent to one side or the other will often give rise to increased hearing, or the reverse. These effects appear to be due to a movement of the mucus in the tympanum, although the precise positions in this cavity which the mucus occupies in good hearing or the reverse have not been made out.

And, indeed, it is difficult to conceive how the matter could be determined during the life of the patient. I should suppose, however, that as the fenestræ are the parts most sensitive, so far as hearing is concerned, to any changes from health, mucus in contact with the internal wall of the tympanum would produce greater deafness than it would in any other situation. Be this as it may, I have on many occasions noticed how a change in position will affect the hearing of persons with the symptoms of mucus in a fluid condition in the tympanum. There is always a reasonable expectation that the hearing will in a great measure be restored so long as this fluid condition remains, but the cases of real difficulty are those in which nothing has been done for many months, perhaps, where the catarrh has subsided, leaving the mucus inspissated, with a thickened condition of the lining membrane of the tympanum. Under these circumstances the hearing remains stationary sometimes for years; it is not affected by inflation, and the sounds heard through the otoscope are of a dry character. The treatment for these cases consists in an endeavour to soften the dried mucus, and for this purpose the injections consist of warm alkaline solutions. After this line of treatment has been pursued for a time it is well to discontinue it for a few months, and again to resume it. When it is considered that the results of treatment depend entirely upon the character, form, and situation which the dried collections have assumed, which it is, of course, impossible to recognise during life, it is not surprising that extensive improvements in hearing are met with in only a fair proportion of these cases.

One of the difficulties which beset us when this condition has been reached is to determine what are the prospects of success in undertaking a course of treatment. These cases, too, are very common, and for the very easily explained reason that persons, especially in the rank of life from which hospital patients come, are content to remain deaf on one side for a long time before they will exert themselves to get advice. By the time they apply to be treated, the period when relief could be given to them has in many instances passed by. You can readily fancy how in a little cavity like the tympanum a small quantity of mucus, permitted to remain until it had become dried up, would clog the movements of the ossicles, and how very slight alterations from health in the way of catarrh of the lining membrane would induce thickening and a consequently less sensitive condition of the fenestræ; and so it is. Case after case will come giving the same history—a complete account of catarrh of one side occurring years and months ago. On examination the tympanic membrane of the deaf side will give corroborative evidence of the character of the affection.

We know then that there has been a catarrhal condition of the tympanic cavity, and we have a general
idea of the state behind the membrane ; the sound
which we hear as air passes into the tympanum tells
us that it is in a dry state, at least that there is no
moist excessive secretion, and this is about all we do
know for certain. Beyond this, any other evidence we
have is purely negative in its nature, and consists in
an absence of symptoms which point to a failure in
the functions of the nervous apparatus.   Thus, if
there is excessive tinnitus, much worse hearing after
fatigue, if a vibrating tuning fork placed on the head
is not heard at all on the affected side, we may be
pretty sure that the deafness is not wholly due to
catarrh.

Neither do I believe can any one tell on the first
examination of such a case as I am describing whether
the patient will receive much or any benefit from
treatment ; so we are forced to almost act in the dark
and be guided by what has been the result of treatment in similar cases ; not in any single case, but the
general success or the failure of any line of treatment
in many.   So far as my experience is worth anything,
I think that the warm alkaline injections most suitable are five grains of bicarbonate of soda, iodide of
potassium, or hydrochlorate of ammonia, to the ounce
of water, or simply warm water.   If you were to take
all the cases that I have treated for the last three or
four years, they would display the greatest discrepancy
as to results.

Sometimes I have had several cases in succession

which have not received any benefit, and following these have been two or three in which the hearing has been very greatly and permanently improved. Putting side by side two individual cases, one of which has been a very successful one, and the other in which the effects of the treatment have been positively nothing, I am not able to account for this in any way. Possibly I have failed to observe some symptom by which I ought to have been led in my prognosis—I cannot say; what I have noticed, however, is that, when the hearing is going to be favorably affected by the injection, there is generally some appreciable change after it has been done a few times; not, as a rule, immediately afterwards (indeed, it not unfrequently increases the deafness for a few hours), but perhaps a day or a few days after the injection has been used. Even in the most successful cases of this kind, when the catarrh dates back for some years, and the secretion in the tympanum has become inspissated, the recovery is, so far as I know, never so complete that the patients regain perfect hearing, neither from the nature of the case is it at all likely that they should do so.

The two most successful results of this kind of treatment in old-standing cases of catarrh that I have seen took place in a man of thirty-six and a boy of twenty. In both instances the first attack of catarrh occurred more than four years before I saw them. In the case of the man there was nothing to be seen beyond a general opacity of the membrane, and there had been some tinnitus for two months. He began to improve after the second injection. The ear was injected with solution of soda, five grains to the ounce, every other day for a fortnight, and twice again after an interval of a fortnight. The boy was treated in the same way for only a fortnight, and made an equally good recovery. In both instances the hearing twelve months afterwards for practical purposes was as good as could be wished, although when carefully tested it was not perfect. Both the cases were very exceptional, and generally under similar circumstances we have to rest content if the hearing can be said to have been decidedly improved.

Before concluding this lecture I must direct your attention to a mode of treatment in cases of tympanic affections more decided in its character than what I have as yet spoken of. You will bear in mind that we have been considering two separate morbid conditions of the tympanum, one in which the mucus is in a fluid or semi-fluid state, and the other where it has become dried.

In the supplement which Dr Millingen wrote to Dr Adam Politzer's work on the membrana tympani, in describing the practice of the last-named surgeon in cases of accumulations in the tympanic cavity, and their immediate dispersion by the air douche, he says, "Secretion is not always so fluid that it can be thus easily removed from the tympanic cavity, for when it has been retained for a long time, a gelatinous viscid matter is formed. This will either be absorbed in course of time or removed by the employment of Politzer's air douche or by the catheter.

"In such cases, when the membrane was much shrunk and presented a dark greenish-yellow colour, Dr Politzer, after using the air douche with only temporary success, performed paracentesis, and immediately afterwards forced air through the ear by this method, thus driving the mucous matter out into the external meatus. The opening in the membrane usually closed by the following day, and the hearing, moreover, was restored to its normal standard or returned after several applications of the air douche."

During the last few years this plan of treatment has been employed in England with varied success, in some instances with considerable advantage. An account of it, with cases, was published by Mr Hinton in the 'Guy's Hospital Reports' for 1869 and 1870. While light is reflected down the speculum from a mirror fastened on the forehead of the operator, a vertical incision, about one eighth of an inch in length, is made either in front or behind the handle of the malleus with a small double-edged knife made for the purpose. If expulsion of the mucus does not

FIG. 11.—KNIFE FOR MAKING INCISION IN THE TYMPANIC MEMBRANE
(*handle in figure half the length*).

follow the passage of air through the cut on Politzer's plan, Mr Hinton advises that a weak solution of soda, or simply warm water, may be passed through the incision by means of a syringe the nozzle of which fits the meatus (it is carefully protected by a rim of india rubber), the fluid passing through the tympanum and

Eustachian tube out at the inferior nares as the head
of the patient is bent downwards.   In this way any
accumulation of mucus in the tympanum is dislodged
from its situation.   The opening made in the membrane is thus merely the preliminary step of the proceeding, and is simply a means to an end, so is in no
way a revival of the operation practised by Sir Astley
Cooper for the relief of Eustachian obstruction and
the troublesome symptom of tinnitus.   This at the
best could only give very temporary benefit, as an
incision at the tympanic membrane heals in from two
to five days.

You will observe, too, that the object of the incision
in these cases is very different from that had in view
when it is made in the course of purulent catarrh of
the middle ear : in the latter instance it is done to
provide an outlet for the pus which fills the cavity of
the tympanum, and which, unless let out in this
manner, causes rupture of the membrane and more or
less disorganization of the contents of the tympanum.

At the same time there is no doubt but that in the
unpurulent variety of catarrh of the tympanum, when
the secretion is very copious, the membrane will some-
time give way in consequence of pressure, and thus an
opportune incision will prevent ulceration taking place.
This, however, is not the principal or only object in
adopting this practice as recommended by Mr Hinton.

A certain proportion of the cases in which it was
done have unquestionably obtained improved hearing
in consequence, whilst others were no better afterwards.   If this can be shown to be the experience of

all who have practised this operation there are certainly quite sufficient data to make the proceeding advisable under appropriate circumstances. Now, what are these circumstances ?

Unfortunately, it is rare to meet with appearances in the tympanic membrane of so distinct a character as to enable us to say with certainty, "there is within that tympanum the remains of catarrh in the form of semi-fluid accumulations of mucus, which can be removed after an incision has been made into the membrane." Such appearances are described, in the report referred to, as spots of a brownish-yellowish colour due to dark-coloured fluid in contact with the inner surface, bulgings of the membrane in places, especially at the posterior and superior part, and thin spots which can be made to bulge on inflating the tympanum.

You will find that in cases where these changes are noticeable the same history is given as in others where there is no more abnormality to be seen than a somewhat exaggerated inward curvature, or a general opacity, which is merely indicative of previous attacks of catarrh. Hence the difficulty in deciding the question as to whether the opening should be made in the membrane or not. The most reliable sign, I believe, which we possess of the presence of fluid mucus in the cavity of the tympanum is the sound which accompanies inflation and is heard through the otoscope. Occasionally, upon the patient's pressing air into the tympanum there is a squeaking and bubbling sound most characteristic of the condition under notice.   It is with such cases that I have never failed to obtain good results after this operation.

My own observations and experience, therefore, go to show that the cases favorable for this treatment are not necessarily either those in which the catarrh is recent or cases of very long standing, but that the condition of the tympanum which I spoke of as a dry condition must not have been reached.

Useful as this operation undoubtedly is in favorable instances, I need, perhaps, hardly say that it is a sufficiently serious one to make it necessary to exercise great care and judgment in the selection of cases, and it should not be resorted to until less severe measures have had a fair trial.   In coming to a decision on this question the occurrence of any of those symptoms which point to a lesion of the labyrinth (which will be detailed in a future lecture) would counter-indicate the operation.

For if the nervous apparatus be faulty it is but lost labour to improve the means by which impressions are conveyed to it.

Although we have so generally to seek for the cause of impaired hearing in morbid conditions behind the tympanic membrane, there is a condition of the structure sometimes to be met with which either directly or indirectly affects the hearing power.   This is when there is a relaxation of the membrane, and, irrespective of any appearance, you can almost diagnose it in the following way.   The patients who are the subject of it have suffered from tympanic disease for many years, and upon putting the membrane on the stretch by inflating the tympanum they can hear for a few seconds pretty well, but as the membrane falls back again they are deaf.

It is, I believe, to the constant and almost habitual habit of doing this that the relaxation must be attributed, and attention must be directed to the primary affection, viz. the tympanic disease, before any good can be expected from treatment. The extremely transient nature of the improved hearing will distinguish the condition from the longer interval of good hearing (at least ten minutes), which is sometimes produced in the same way when the deafness is caused by obstruction of the Eustachian tube. On examination of these cases occasionally the exaggerated outward movement of the membrane following inflation can be clearly seen.

This appearance is occasionally of so marked a character, that as the air is forced into the tympanum, the relaxed and thinned portions of the membrane will be forced outwards until small bladder-like protrusions may be seen. Sometimes one of these protrusions will (in consequence of the very thinned state of the part of the membrane which forms it) give way, and the case is at once converted into one of perforation, for, the external air being admitted into the tympanum, the lining membrane of this cavity becomes a suppurating surface; in other words, the opening is a fistulous one. In order to obviate this I have occasionally had recourse to the following device. The patient is taught to make a small, flattened disc of moistened cotton wool, and to adjust this on to the membrane; the support thus formed is worn through-out the whole or a part of the day. After this plan has been pursued for some time it is not uncommon for the membrane to regain its tenacity, and thus the ligamentous support which it habitually gives to the malleus is restored.

As might be expected, from time to time many attempts have been made to establish a more or less permanent opening in the tympanic membrane in cases of long-standing chronic catarrh, for it has been found that better hearing has frequently followed an artificial orifice in this structure, but has not been maintained for more than a few days, owing to the ready way in which incisions in the membrane heal. Small pieces have been cut out, but with a similar result. The nearest approach to success in this direction has been achieved by Dr Politzer, who, after making an opening

FIG. 12.—POLITZER'S EYELET.

with a knife in the posterior section of the membrane and dilating it with a small laminaria tent, introduced a little eyelet made of hard rubber, and having a groove in which the edges of the cut membrane rested, and thus held it in position. Into the eylet is fixed a piece of silk which lies in the meatus, so that there is no fear if the little instrument slipped into the cavity of the tympanum, of giving the trouble of recapturing it. In adopting this proceeding I have not found it necessary to dilate the incision, but have at once pressed the eyelet into the orifice with the forceps adapted for the purpose. In a good many cases the eyelet will remain for several weeks, and occasionally for months, in the position in which it has been placed, but it will sometimes be necessary to insert a fresh one, as the first more often than not slips out after a short time.

After the eyelet has been finally removed the small artificial perforation very rapidly heals, and it is this general tendency of the tympanic membrane to heal which makes all operations of this kind so uncertain in their permanent effects. I must not omit to mention an operation which has recently been practised by Dr Weber, of Berlin, in cases of old-standing tympanic catarrh. I refer to division of the tensor tympanic muscle. Whenever there has been obstruction in the Eustachian tube or tympanum we have seen that the tympanic membrane becomes indrawn, the ossicles rotated more than usually inwards, and consequently the stapes unduly pressed against the fenestra ovalis. As a consequence of this, there will be intra-labyrinthine tension, and from the position assumed by the tympanic membrane it will follow that the tensor tympani may become subject to permanent shortening. The object in view in division of this muscle is plain, viz. to set free as far as possible the tympanic membrane from the effects produced by muscular contraction, and thus indirectly relieve tension within the labyrinth.

More experience is needed before the conditions favorable for this operation can be determined. At present, however, it may be said that in some instances considerable benefit is reported to have followed its practice. It is stated that the most noticeable direction in which benefit has been derived from this treatment has been diminution of the tinnitus.

It is, comparatively speaking, a rare circumstance for the effused secretion in the tympanic cavity to remain for any lengthy period in a fluid condition, and it is still more rare to meet with cases in which the fluid state can be demonstrated by sight. I have only seen one instance of this, and the case is reported at some length in the 'Practitioner' for April, 1872. From the very clear account given by the patient the tympanum had been the seat of this change for at any rate three weeks, and possibly for much longer. The tympanic membrane retained a perfectly healthy appearance. While the patient inflated the tympanum on the Valsalvan method there was the "most distinct appearance of fluid running slowly down the interior surface of the membrane. I can compare it to nothing better than drops of water collecting and running down a pane of glass that is being rained upon. After using Politzer's inflation the fluid seemed to be dispersed into drops and bubbles, and then to collect and roll down."

A remarkable appearance of this kind has been observed before by Politzer, and he speaks of the case as an accumulation of serum in the tympanic cavity. Instances of this kind may be regarded as merely unusual phases of catarrh of the tympanum, and their treatment demands no special notice. The patient I referred to recovered completely after the air douche, though the catheter had been used for only six times. In very obstinate cases it may possibly be found necessary to make an incision into the membrane and thus bring about the evacuation of the fluid.

You may, perhaps, be surprised that in all I have said on the subject of treatment no mention has been made of blisters as applied over the mastoid process. If I shared the opinion, still retained, I believe, by some, that the chief available resources for treating diseases of the ear consisted in the use of the syringe and blisters, I should have hesitated before I criticised unfavorably one of these agents, lest I should haplessly leave you with only one to fall back upon.   An exaggerated idea of the potency of blisters behind the ear is not altogether to be wondered at when we remember that it has been a time-honoured practice to advise this application for all acute, chronic, inflammatory, and non-inflammatory affections of the external, middle and internal ear.

So common is it even now, that it is rare to meet with a patient suffering from disease of the ear of any long standing who has not, at one time or another, been subjected to this favourite, and fortunately harmless remedy.

In an article on counter-irritation by Dr Dickinson in the ' St George's Hospital Reports ' for 1868 the empirical manner in which blisters are ordered by practitioners of successive generations is graphically shown.   Wholesale blistering behind the ear must be added to the list of examples there recorded, where counter-irritation is prescribed in a truly irrational manner. This agent in treating ear affections has long been discarded altogether in Germany, and it is desirable that the example thus set us should either be followed, or that it should be determined in what class of cases blisters are likely to be serviceable.

If at any time we can discover a class of cases where they unquestionably have a good effect, even though we cannot explain the way in which they act, it were better to prescribe blisters empirically than not at all. To have our faith shaken in any belief is a distressing process, and in the question before us we must remember that Mr Toynbee, in his book ' On Diseases of the Ear,' recommends again and again blistering over the mastoid process and the administration of alteratives, and this with his very large experience. I have some hesitation, therefore, in saying that, from what I have seen, it seems to me highly probable that beyond a disagreeable and somewhat disfiguring sore behind the ear, which serves to remind the patient of the trouble for which they have been applied, they produce no effect on the majority of cases in which they are employed. They may possibly be useful in the extremely uncommon condition of serum in the cavity of the tympanum, and this on the principle that the vesicated surface has a vascular connexion with the cavity to be acted upon. In this way, the withdrawal of serum by means of blisters may be expected to excite the resumption of the effused serum by the blood-vessels. This is probably the reason why these appear to be useful in cases where the facial nerve has

been partially paralysed after an attack of inflammation of the tympanum, for you will remember that
the portio dura passes through the cavity and is protected by a thin plate of bone, which forms the
aqueduct of Fallopius.   This plate of bone is not only
very thin, but is sometimes perforated by small holes.
When in this state the nerve is more than usually
exposed, and so suffers from inflammation, which need
not necessarily be so severe as to destroy its bony
protection.

LECTURE V

LET us briefly review the difficulties which present themselves in forming a diagnosis of the condition of the middle ear in non-purulent catarrh, and giving an opinion as to the results to be expected from treatment. We have seen that simple Eustachian obstruction which has not involved the tympanum is alike easy of diagnosis and treatment, and that the subjects of this will get well if treated within reasonable time. That when the affection has extended to the whole of the middle ear, so long as it is not of very long standing, recovery may be expected, and that the treatment consists in the main in the air douche regularly applied with Politzer's inflation, or the Eustachian catheter, and astringent solutions injected to the lining membrane throughout the middle ear. That if this unhealthy condition of the middle ear be allowed to exist unchecked for some time, as in the case of some other morbid processes, it may arrive at a state of quiescence, and that certain pathological products, having their seat in the tympanum, result in consequence, which products, having assumed various forms in this situation, will so act as to become impediments to the passage of sound to the labyrinth. That the forms which these products take are most varied;

that it is frequently with the greatest difficulty they can be estimated during life, but that upon these forms depend the answer to the question whether the disease is remediable or not ? Lastly, that we must at times be content to act on certain general principles to such cases, in the confident expectation that in whatever proportion we can soften or remove these results of disease, which are obstructing the passage of sound through the tympanum, so will the hearing power improve.

The affection of the tympanum next in order is that in which inflammation of the mucous membrane has proceeded to the formation of pus : hence the term commonly applied to it—namely, acute purulent catarrh. The most usual circumstance under which this occurs is an attack of some one of the exanthemata, most commonly scarlet fever, the inflammation having spread up the Eustachian tube from the throat. It is also met with independently of these diseases, and often cannot be traced to any exciting cause, at least so far as can be discovered. In this case the inflammatory action begins in the tympanum. It commences with an earache ; the pain increases in severity until it is of the most agonising and sickening character. General feverish symptoms soon set in. On · examination of the tympanic membrane, there is in the early stage little change to be seen beyond congestion of the vessels, and the external meatus is not affected. Inflation of the tympanum aggravates the pain, and there is more or less deafness. This latter symptom varies very much. Some patients are

extremely deaf, and others very slightly so. There is
nearly always tinnitus, and sometimes it is severe.
The act of swallowing increases the pain, as does also
any movement of the head. If allowed to proceed
unchecked, in from forty-eight hours to three or four
days the membrane gives way, a discharge of pus
appears, and the tension being thus relieved the pain
ceases. Such is the ordinary course of an acute attack
of inflammation of the tympanum. In scarlet fever
the same process goes on; but, on account of the more
serious symptoms which endanger life, engrossing the
attention of the friends (this especially with children),
the symptoms referred to the ear are not regarded;
and it appears, although no explanation has been offered
for the fact, that, when accompanying the exanthemata, inflammation of the tympanum does not give rise
to so much pain as in the idiopathic variety. It does
not invariably happen that all cases of this kind even-
tuate in a perforation of the membrane. The pus may
pass into the pharynx through the Eustachian tube;
but this mode of escape is very seldom practicable, as
there is so much swelling of the lining membrane of
the tympanum, that the tympanic opening of the Eustachian tube becomes obstructed. If a case is seen in
the early stage before the formation of pus, the attack
may be cut short by putting on leeches in front of and
below the ear, followed by fomentations. The pain
will sometimes subside after this, and the patients
recover. If this does not happen, the injected condition of the vessels on the tympanic membrane is succeeded by a sodden appearance. Before the membrane

gives way a distinct bulging in of some part of the membrane may sometimes be detected. Whenever this can be seen no time should be lost in giving free vent to the pus by making an incision in this situation; and even if this distinctive appearance is not observable, so long as there is sufficient evidence of purulent matter in the tympanic cavity, there can be no question of the propriety of making an opening. The incision should be about two lines long from above downwards, and may be made with the little straight-bladed knife which I showed you at the last lecture. Of course at the time the mirror will be worn upon the forehead, and light be thus directed down the speculum, so that there is no difficulty about doing it. If the membrane does not show symptoms of giving way in one particular spot, about the middle of the posterior section, and in a line with the handle of the malleus, is the position generally selected for the incision. After this, fomentations should be applied, and the ear syringed frequently with warm water. To thus provide an exit for purulent matter in these cases is demanded by the recognised rules of surgery which guide us in similar instances; and if an examination of the tympanic membrane were more usually made, disorganisation of the tympanum and ulceration of the membrane causing partial or complete deafness, besides the later occurrence of cerebral abscess and secondary deposits, would be more uncommon events than they are now.

As the external or dermoid layer of the tympanic membrane will then be the first to become involved in the course of inflammation of the external auditory meatus, and the internal or mucous layer in the course of inflammation of the cavity of the tympanum, so the one or the other may proceed inwards or outwards, thus involving the middle or proper fibrous layer in its course. It is extremely unusual for the membrane to be primarily affected, and the disease spoken of as myringitis is for this reason, I believe, much rarer than what might be supposed from the accounts that have been written of an affection to which this name is applied. A case to which this term would perhaps apply occurred in a boy whom I saw a few hours after he was taken with pain in the ear. The tympanic membrane was at that time injected, so that the filled vessels could be traced out in their course; the next day a very network of vessels, so that it looked almost purple; there was great tinnitus, acute pain, increased on inflating the tympanum; he was scarcely at all deaf. Leeches and fomentations formed the treatment, and in a week all unpleasant symptoms had gone, the handle of the malleus was again visible, and, excepting an absence of lustre, there was nothing unusual in the appearance of the tympanic membrane.

A very large number of the patients who come to the hospital for advice in the department for diseases of the ear are the subjects of a purulent discharge from either one or both ears. So common a symptom is this, that among the questions which are asked them in the ordinary routine, before an examination is made, is the one of—Have you ever had a discharge from the ear?

In case of any patient suffering from this symptom, which is spoken of by some writers as otorrhœa, in most instances it will be found to be accompanied by a perforation in the tympanic membrane. If this can be seen upon inspection, or if the patient can, with the mouth and nose closed, blow through the external meatus, or if, on syringing the ear, the water passes into the throat, the evidence is conclusive. Any one of these methods of examination may, however, for various reasons to be mentioned further on, fail in demonstrating a perforation, even if one be present, so that it will be necessary not to rely too completely on them individually. For example, it is sometimes impossible to detect a small perforation in the anterior and inferior part of the membrane. Some adults do not readily learn how to pass air through the Eustachian tube ; and without the hole in the membrane be very large, in syringing the water does not pass down the Eustachian tube ; and, again, the passage into the pharynx may be obstructed.

Politzer's method will overcome any mere obstruction, and the air will pass out of the perforation with a squeaking sound that can hardly be mistaken. If the Eustachian tube be occluded this method for diagnosis is not available ; but under these circumstances the perforation is generally large and of long standing, so that it can be easily seen. Even if its position is one that can be readily seen, occasionally, where the loss of tissue has been very slight, a perforation may escape notice. In such a case as this, if while the membrane is under observation, the patient be directed to blow through the rupture, the sides of the perforation will become separated, and on this movement being suspended, will be observed to fall back again into position.

An appearance that is not at all uncommon in these cases, and one that was originally pointed out by Sir William Wilde, is a pulsation taking place synchronously with that of the arteries.   This it would appear is dependent on a drop of fluid being in contact with a blood-vessel on the swollen mucous membrane lining the tympanum, for it will disappear sometimes when the spot is carefully dried.   The presence of fluid, however, is not always necessary for this appearance to be observed, and it will therefore occasionally be constant.   Having established without doubt the presence of perforation, the next question is to what does it owe its existence ?

Of all causes of perforations, the purulent catarrh of scarlet fever is the most frequent.   The extent of disorganisation which the cavity and contents of the tympanum undergo will in a great degree be proportionate to the length of time which the catarrh may have lasted before the purulent matter has found an exit.   If this event be long delayed, the ossicles may all come away ; there may be necrosis of the petrous part of the temporal bone ; from time to time pieces of dead bone which in their entirety represented the labyrinth will become loose in the meatus, and may be drawn out ; all this of course producing total deafness, and in children being a fruitful source of deaf-mutism.   The sooner, therefore, the tympanic membrane gives way, if it does so at all, the better; for when this takes place early in the course of the affection there will be less change wrought in the parts behind this structure. Less frequently than scarlet fever as a source of perforations are the other exanthemata, measles and erysipelas ranking perhaps highest in the list. Without any assignable cause, infants are the subjects of purulent catarrh of the tympana, which passes unnoticed until shown by a discharge from the ear, and even then does not seem to give much anxiety to the parents if we are to judge by the casual way in which this symptom is frequently mentioned as having gone on for years. There is little doubt, however, that the inflammation in the case of these little patients begins in the cavity of the tympanum, and they suffer from earache and cry, but are unable to direct attention to the seat of pain. This infantile tympanitis has been pointed out by Mr Hinton as being in all probability sometimes the unrecognised cause of convulsions. The extreme pain of the affection, the close proximity of the brain to the affected part, would be very much in favour of such an hypothesis.

With patients who have previously had attacks of non-purulent catarrh slight exciting causes are sufficient to lead to a rupture, as in their case the membrane has been the seat of various changes which have rendered it especially prone to give way under circumstances which in health would have been very inefficient to produce this lesion. It may be said generally that perforations owe their origin to causes situated behind the membrane. Among the exceptions to this rule are accidental perforations, and those arising from diffuse inflammation of the external auditory meatus. This latter is much rarer than is generally supposed; still more rare is what has been stated by some observers to result in a perforation—viz. minute abscesses situated in the substance of the membrane itself.

Do not suppose that in every case of a perforation we can obtain a definite history to account for it. It is often impossible to get any history at all. Patients will sometimes say that they have gradually become deaf, and that a few days before they apply for relief they noticed a slight moisture at the external orifice, perhaps preceded by a feeling of uneasiness. Badly fed, unhealthy, tuberculous patients are sometimes in this case; and occasionally in the last stage of phthisis, a few days or hours before death, the tympanitic membrane gives way.

Scalding will occasionally give rise to a perforation, I have seen several instances of this. The last one here was in September, 1871, when a teapot of hot water was spilled over a child's head. The introduction of irritating lotions prescribed by quacks must not be omitted. I was once consulted by an apparently intelligent person, who for a temporary deafness which, by the way, afterwards became permanent, had been induced to put urine into the ear; the effect of it was inflammation, and subsequently perforation, of the membrane. These, however, should come under the category of perforations the result of accident, or rather of folly.

Having ascertained, then, as well as we can the

causes of the loss of continuity of the tympanic membrane, on proceeding to examine a number of cases we shall find that the appearances present great variations as to size, shape, and position; while some are nothing more that a small circular hole, as if made by a pierce from a needle, almost complete absence of the membrane will be seen in others—not total destruction, as there will generally be a rim, however slight, which marks the position previously occupied by the healthy structure. Between these two degrees " an infinite variety."

But, notwithstanding this, in examining a large number of patients with perforations we cannot fail to be struck with the similarity which often one case will bear to another. I mean, we can at times say on looking at one patient, this is very like the appearance in so and so's ear; thus there will be typical perforations as well as typical disease of any other kind. It is almost unnecessary to say that the ear should be carefully syringed before anything at all can be made out in looking at a case for the first time where discharge has been allowed to go on for a long time without anything being done for it.

In the largest forms of perforation, that part of the meatus close to the membrane is very red and swollen, with its granular surface continuous with the cavity of the tympanum, the mucous membrane lining this in like manner so red and swollen that it is impossible at first sight to distinguish any individual part, the whole looking like a cup-shaped cavity. On closely examining it a rounded prominence may be distinguished,

the promontory; in front of this a depression indicating the direction which would lead into the Eustachian
tube.   This is about all that can be made out.   If,
however, under treatment, the spongy condition of the
mucous membrane become more healthy, the upper
part of the malleus may be detected (the handle in
these cases is generally gone).   It is not very uncommon where the malleus retains its position to find
the posterior section nearly all gone, and the posterior
border of the malleus is then free, and the anterior
section of the membrane remaining; or, again, triangular pieces of membrane may be absent, the apex of
the triangle being represented by the umbo.

Whenever the support which the tympanic membrane gives to the malleus is removed by part or all of
the membrane being lost from ulceration, the handle
of this bone becomes drawn upon by the tensor tympani, so that the handle is never seen by itself dividing
the opening into the tympanum in two halves as you
might expect, but becomes tilted inwards, so that it is
difficult to recognise it; but more generally than not
where the loss of the membrane has been very extensive either the handle becomes detached or the whole
bone is lost.

Again, where the destruction of the membrane has
been very complete, and the malleus and incus have
come away, it occasionally happens that the position of
the fenestra rotunda, and the stapes or rather a part of
it may be seen, the first being behind, and the second
above the promontory.   Another common appearance
is when the perforation has the shape of a kidney with

perhaps clean sharp edges, the extremity of the handle
of the malleus marking, so to speak, the hilus of the
kidney.   A slit-like opening, as if produced by an incision with a knife, or a little round hole, as if punched
out of the membrane, either in front or behind the
handle of the malleus, are also among the forms chiefly
met with.

The experience of Sir William Wilde would lead him
to think that the anterior section of the membrane is
the more usual situation for a perforation than the
posterior, and this because (he says) of the pressure of
air from the Eustachian tube falling directly on to
this part.   For myself I have not noticed in those
cases where the handle of the malleus has remained
fixed in the membrane that the perforation is more frequently either anterior or posterior to this boundary
line.   The edges of a small perforation will be sometimes sharply and cleanly defined, at others rugged,
with a red upraised edge, and the general appearance
of a perforation will alter very considerably during the
course of treatment.   You can, however, fancy how
multitudinous are the appearances, when at one time
part of the malleus, at another, the whole of this bone,
at another, part or the whole of the incus as well, and
sometimes the stapes, have come away.   Added to
this, that what remains of the tympanic membrane
may, in part or altogether, have contracted adhesions
to the walls of the tympanum, for so soon as the support which in health the malleus affords to the membrane is withdrawn, there is often a complete or partial
collapse of the remaining portion of the membrane.

Instead of describing now all the appearances that are met with, it will be more useful to observe them as they actually come under our notice in the cases. I may add, however, that I have on several occasions observed a double perforation—by this I mean two small perforations in the same membrane—each time both peforations were in the same section of the membrane, *i.e.* either in front of, or behind, the malleus.

Leaving to a future lecture the consideration of the fatal effects succeeding a perforation in consequence of the brain becoming affected, or pyæmic deposits taking place, I may say that it is for the discharge from the ear and the accompanying deafness that patients apply for relief.

In the course of scarlet fever, while suppuration is going on in the cavity of the tympanum, the labyrinth may be implicated, and in a few hours total deafness may result. If the bone become diseased, and the part forming the labyrinth necrose and come away, the same effect will of course follow, and both states are, I need not say, irremediable. Generally, however, the deafness is not extreme, that is, not so much as to require a very loud voice close to the ear. The extent of this symptom cannot be measured by the size of the perforation. Very fair hearing will be found with almost total loss of the membrane, and very considerable deafness with very small perforations. Indeed, the amount of tissue lost has, I have convinced myself, very little to do with the amount of hearing lost. On this point Dr Adam Politzer says, " The patient hears better as a rule when the perforations are of medium size than when they are very small, because in the former case the sonorous waves, avoiding the membrana tympani, malleus, and incus, pass through the perforation directly upon the base of the stapes, and may thus reach the labyrinth to quite an extent." And Sir W. Wilde says, "I have observed that when once the tympanal membrane has become permanently open, the larger the aperture the greater the amount of hearing, provided no further mischief has taken place, and that there is a slight ring or circle of the membrane still remaining."

The loss of hearing, I believe, almost entirely depends on the effects which the suppurative process has produced within the tympanum. The whole lining membrane of this cavity is altered; there may be anchylosis in any part of the chain of ossicles, or of the stapes into the fenestra ovalis (this produces very great deafness), absence of any part of or all the ossicles, the formation of adhesions connecting them to the walls of the tympanum, thickening of the membrane covering the fenestra rotunda, and a variety of other changes, all of which have been found on examination after death.

The constant discharge of pus (secreted from the lining membrane of the tympanum) through the perforation keeps it open, and if the secretion in the tympanum ceases, perforations occasionally heal. The first consideration in the treatment of these cases should be to induce a healthy condition of the mucous membrane throughout the middle ear. This result may be ac-

complished in two ways. First, by the application of
appropriate solutions to the whole of the lining membrane of the middle ear, and secondly by the protection
of the tympanic cavity from the external air. Before
the application of any solution to the surface referred
to, it is a most necessary preliminary step that the ear
should be thoroughly cleansed (by syringing) of all
accumulation of purulent matter. In order to accomplish this it is not only important that the patients
should be supplied with a convenient form of syringe
and taught how to use it, but they should also be
instructed how to blow through the perforation whilst
the stream of water is injected into the ear ; for in this
way the accumulations of pus and general *débris*
are expelled from the meatus.

Of course, in the case of children, the attendant
must have charge of the syringe, and very young
children are not readily taught how to blow through
the ear. The same proceeding, viz. blowing through
the perforation, must be adopted when the lotions
that are used are placed in the meatus. As the fluid
is thus made to bubble in the ear, if the blowing
be suspended for a moment some of it will pass
down the Eustachian tube into the throat, and be
known to have done so by the taste. You may be
sure, unless this takes place, that the remedies, whatever they may be, have not been properly applied. At
one time I, in common with most other surgeons, was
in the habit of ordering solutions containing some
mineral astringent such as sulphate of zinc. For
many years past I have discontinued this practice, as ex-

perience has taught me that mineral astringents are not advisable, for although their employment will be found for a time to diminish the discharge, it will return in a short time after they are discontinued.   Moreover, the hearing power does not receive permanent benefit from such treatment ; indeed, the reverse of this is the case.   Amongst the best kind of applications will be found weak solutions of alcohol.   These should be used at first very dilute, such as one drachm of spirit to a third of a tumbler full of warm water, and the strength may be gradually increased until a slight burning sensation in the ear follows the injection. Anything stronger than this, *i.e.* enough to cause pain, will be too strong.   Some patients will in time bear as much as one part of spirit to the other of hot water, but not many.

In the next lecture I shall speak of the second and more important part of the treatment of perforations, viz. the protection of the tympanic cavity from the external air.

Another plan of applying astringents to the lining membrane of the middle ear was introduced by Dr J. Gruber, of Vienna.   The patient bends his head to the affected side, he then with a small syringe injects some of the solution into the lower nasal meatus of that side, and, holding the nostrils firmly between his fingers and keeping the mouth closed, blows the solution into the tympanum and through the perforation.   This is not so complete and cleansing a process as the other in which the solution is passed in a contrary direction ; it requires more intelligence and trouble on the part of the patient than is always met with, and is inapplicable for children.

If there is a temporary obstruction of the Eustachian tube, it may be overcome sometimes by injecting a little water through the Eustachian catheter if it does not yield to the air douche.

LECTURE VI

In the last lecture we considered the general principles of treatment in cases of perforation of the tympanic membrane. There are some other matters connected with these cases which remain to be noticed.

It is not necessary here to refer to the history of the introduction of the so-termed artificial membrane by Mr Toynbee and Dr Yearsley respectively. An account of the one can be seen in Mr Toynbee's work on the ear, and of the other in the 'Lancet' for July 1st, 1848. The one consists of an india-rubber

FIG. 13.—THE ARTIFICIAL MEMBRANA TYMPANI (*Toynbee*)

disc fixed to a piece of silver wire, and worn in the position occupied by the tympanic membrane; the other of a small plug of cotton wool moistened with water or glycerine, and adjusted by the patient with the help of a pair of forceps to the same spots. The latter form is the more simple of the two, and when it

FIG. 14.—FORCEPS FOR ADJUSTING COTTON WOOL.

produces equal improvements in hearing to what is done by the other, it is by far the more preferable, its tendency being for good on the exposed surface of the tympanum, while the effects of the india-rubber disc is not unfrequently irritating, and increases the discharge. Judging only by the appearances of a perforation, no amount of experience can detect for certain whether a case will be benefited by any form of artificial support. In each one the effect of applying it should be noted. After a few attempts the patient soon learns to adjust it, and, when he has had a little practice, can direct it to the exact spot requiring pressure far more readily than any one can do it for him. The effects of this mechanical aid to hearing do not arise from supplying the place of the natural membrane as a surface on which to receive and through which to communicate vibrations of sound to the labyrinth, nor, as Mr Toynbee at first supposed, by confining the vibrations of sound to the tympanic cavity, but by exerting the requisite pressure on the stapes, and so on to the fenestra ovalis. So long as this be exerted, it were better, so far as hearing is concerned, for the perforation not to be closed artificially, but to allow sonorous vibrations to pass directly through the tympanum ; for a piece of india rubber or a plug of cotton wool is the reverse of susceptible to vibrations of sound, and, as far as a conducting medium is concerned, is rather in the way than otherwise. In cases of perforation the normal pressure of the stapes becomes altered, as the support which the tympanic membrane in health gives to the ossicles is more or less taken away; and in cases where the malleus is gone it is completely so. The increased hearing may therefore be said to be produced by approximating the articulations of the ossicles, or supplying their place when these bones are partially wanting.

This was in substance the explanation offered by Dr Yearsley, and subsequent experience has proved it to be a correct one.

There are a certain number of cases in which the india-rubber disc will produce good effects on the hearing when the moistened cotton fails, and *vice versâ*; but generally if the one succeeds the other does so as well. Although I have known Toynbee's artificial membrane to be worn for years, it very often gives rise to great discomfort, and sometimes is intolerable to the patient, so that whenever the moistened wool effects good hearing I always recommend it.

In applying Toynbee's artificial membrane, it is necessary to cut the disc as near as possible to the size of the natural membrane, moisten it with warm water and gently press it down the meatus; when it has arrived at the right place there is generally a little click, and it comes to a stop. The sensations of the patients are also a guide in this respect, as, if it is successful, they instantly hear better. Great gentleness should be used in introducing this or the cotton wool.

In the latter case the pledget of wool should not be large enough to block up the meatus ; there ought to be plenty of room between the wool and the walls of the meatus for sound to pass to the tympanum. If by these means you can produce good hearing, the patients, be sure, can do as well and (as I said before) better, so that they must take the trouble to learn, and an intelligent person will not be very long in doing so. Instances out of number might be recorded where the subjects of perforation of the tympanic membrane, by using this simple contrivance, are enabled to hear quite sufficiently for ordinary purposes, and when not wearing it are quite useless for conversation.

I must here warn you not to be too hasty in giving a decided opinion as to whether the hearing is likely to be improved by the cotton wool. If you do not succeed in making the patient hear on the first attempt you may do so subsequently, and pressure should be made at the same sitting several times. If you too hastily say of a case, "This will not improve with cotton wool," it is not altogether impossible that you may find some one else has succeeded where you have failed, and this because you, perhaps, in your haste not having exerted pressure on the precise spot required on the first attempt, have not tried again.

I may here say that so far as the shape of the cotton-wool support is concerned, there is room for very great ingenuity, not only as to the size and shape that may be most advisable, but also as to manner in which pressure may be applied to the tympanum in such a way that the normal pressure on the stapes may be artificially imitated. Nothing short of the most patient and painstaking practice will give you skill in this matter.

The principles, therefore, that should guide us in treating all cases of perforation when the conducting part of the auditory apparatus only is at fault, that is, when the labyrinth remains unaffected, may be summed up by saying that the cavity of the tympanum should be kept free from secretion, that the lining membrane of the middle ear should so far as possible be restored to health, and that the artificial support of which I have spoken should be tried. This routine will be found to apply, however long the patients may have suffered, or however recent the affection, and this irrespective of the amount of tissue destroyed by ulceration. Generally speaking, I find that a perforation of large or moderate size receives greater benefit from treatment than a very small one, and this I attribute chiefly to the fact that in the former cases we can more easily treat the diseased part behind the membrane. In some remarks I made on the subject in the 'Lancet' of August 20th, 1870, is the following sentence which I will read :—" When the disease appears to be tympanic, but when after treatment the condition of the lining membrane of the tympanum and Eustachian tube has become satisfactory, without a corresponding change in the hearing, and when the artificial membrane of cotton wool has no effect, I cannot but think, that if the history and symptoms be carefully investigated, there will generally be found a sufficient explanation in the co-existence of a nervous affection with the perforation and tympanic disease ; and I would urge the importance of recognising this condition, as it is only by so doing that those cases which admit of treatment and those which do not can be properly separated." A sketch of a pronounced case of this kind will give an illustration.

A patient of from thirty to forty or fifty years of age, who is the subject of perforations which date for many years back, perhaps from childhood, has had moderately good hearing, with very slight variations, until within a few weeks of being seen. He may tell you that two months previously he became more deaf than usual, and that now his deafness is extreme; he has occasional fits of giddiness and constant noise in both ears; is getting worse. A vibrating tuning-fork placed on the head is not heard at all. It requires very little questioning to recognise in this state of things an affection of the labyrinth of a kind which might have been present without any disease of the tympanum, and it is, of course, irremediable. In other words there is an affection of the nervous structures in addition to that of the conducting apparatus. This is an exaggerated case, but in a minor degree any one or all of these symptoms may be discovered, and will account for the failure in the ordinary treatment. When we experience a disappointment in the effects of our remedies we must look for something beyond the middle ear. Try how the vibrations of a tuning-fork are heard on the head ; they should be heard as well with a perforate as an imperforate membrane. Put the questions, "Are you more deaf after fatigue or excitement?" and "Have you been getting steadily worse since any particular date?" If local causes (by this I mean what are situate in the middle ear) will not account for such a change it is a bad sign. Tinnitus, especially severe tinnitus coming on and remaining constant, with periodical attacks of giddiness are among the unfavorable symptoms which will elucidate the painful nature of the case.

Questioning of this kind should in every case be put when a patient is seen for the first time, and the responses will influence any opinion that may be formed as to the chances of recovery from the impaired hearing.

The exposed surface of the tympanum, when there has been a perforation of long standing, will often be covered with exuberant granulations. This state is not confined to the cavity of the tympanum, but it is not unfrequently seen on the surface of what remains of the membrane; also when the meatus has been the seat of inflammation I have seen the tympanic membrane unperforated and covered in the same way. Whatever the position of the granulations the same treatment will be found most useful.

After the ear has been syringed a camel's hair brush is dipped in the strong Liquor Plumbi or on powdered gallic acid and applied freely to the part. By these means, in a very short time, the granulations disappear and the surface becomes more healthy. A strong solution of nitrate of silver ʒj to the ounce applied on the brush in the same way has also very good effect, but the patients should not be trusted with this as they cannot apply it without painting the whole meatus with it each time, and this, of course, is not desirable. It is better that you apply this through a speculum while the reflector is worn on the forehead. In using the solid nitrate of silver it is convenient to melt a little in a porcelain dish and dip the end of a probe into it so that when it cools, the probe thus armed does for a caustic holder.

In the course of treatment of the perforations, it is necessary when it exists to relieve obstruction of the Eustachian tube either by Politzer's method or with the catheter; but if there is occlusion, neither of these plans, of course, is likely to be useful. By "occlusion" is meant, where cicatricial tissue completely closes the tube; the position of this being the tympanic orifice. This condition is never present unless there has been great disorganisation of the tympanum; and the extreme deafness in these cases in all probability depends on this, rather than on the closed tube, and, as a rule, artificial support does not do much good (this is, however, subject to occasional exceptions).

One case especially, in the early part of 1870.—I examined the ears of a gentleman, thirty-two years of age, who had an occluded Eustachian tube and a large perforation of the tympanic membrane on the left side, the result of scarlet fever in childhood. He had worn the cotton wool for ten years and could with it hear conversation extremely well, but without it could not distinguish a word, even if spoken close to the ear. It was a fortunate circumstance for him, for he was on the right side totally deaf. As a rule, with these cases, all we can do is to check the discharge and induce a healthy state of the tympanum. Even if no more is effected than this, it is by no means a small boon for a person who has suffered for a long time, years perhaps, from a discharge from the ear to be free from it; and, putting out of consideration the comfort of the patient, it should never be forgotten that any person with a perforation of the tympanic membrane and a discharge from the ear, if no attention be paid to it, is more or less in danger of losing his life. It is but a very thin plate of bone that separates (as you know) the tympanum from the interior of the skull. Suppuration in the tympanic cavity may at any time give rise to caries of this septum, the effect of this being meningitis, abscess in the substance of the brain, or purulent deposits from pyæmia. It is, therefore, no less necessary to attend to the discharge than to the deafness. Another serious and a not uncommon complication is when the purulent matter in the tympanum passes into the mastoid cells, or, to speak more correctly, when the lining membrane of the mastoid cells becomes the seat of inflammation. This if neglected may lead to a fatal result, but I hope to show that with proper and well-directed attention such a conclusion to cases of this sort may be avoided. A most excellent account of disease of the mastoid process with its effects will be found in Mr. Toynbee's book on diseases of the ear. He drew attention to what you can see for yourselves on comparing the temporal bone of child under twelve months of age and one from an adult, that in the former case the mastoid process is imperfectly developed, and that the cells in this part are on a level with the tympanum, that a continuation of the same plate of bone separates that cavity and the mastoid cells from the interior of the skull; and that, therefore, in the infant if disease proceeds inwards, or rather upwards, from the mastoid cells the dura mater and the cerebrum are the parts that become affected. After two or three years, however, the bony septum is much thicker, and the mastoid process and cells become developed inferiorly and posteriorly, so that in a temporal bone of an adult if we break into the mastoid cells from without and hold the bone up to the light we shall see that the cells are separated from the lateral sulcus inside the skull, and, therefore, on disease proceeding inwards the lateral sinus and cerebellum will, in adult life, be the parts involved.

As in another lecture I shall speak of fatal cases I say no more on this subject at present; but to look back to an earlier part of the history of these patients, whether it be a child or an adult with a perforation of the membrane, where there is pain, swelling, redness, and pitting on pressure over the mastoid process, a free incision should be made without delay down to the bone about an inch from where the ear joins the head. Supposing that this has not been done, that the pus has made for itself a mode of exit, and that there is an open wound behind the ear discharging pus, I have known patients to have such a wound, off and on, for ten and twelve years together; sometimes it closes up and again breaks out. If the syringe be used with the nozzle blunt-pointed, covered with india rubber, and fitting the meatus, water will be made to pass through the tympanum and out of the opening behind the ear, very generally in a full stream, bringing away in its course any purulent matter and caseous deposits that may be at the time lying in the tympanum and mastoid cells. This should be repeated daily, and after it has been done for some time its beneficial effects will be apparent, for before long the discharge will become less and less until it ceases, and the wound will heal soundly.

On July 20th, 1870, I adopted this method of treating a case in a boy, J. G—, of 19, who had had a discharge from the ear since infancy, and an opening with discharge over the mastoid process for six months; by 20th August the wound had healed.

In November of the same year a girl of 20, M. M. W—, who had had an open wound over the mastoid process for five years recovered completely after three weeks' treatment. I saw her in the following November, 1871, and she was still quite well. This is a very simple and, I submit, a very rational and successful way of managing what are generally very troublesome cases, and I commend it to your notice. I mention these two cases especially as they were some of the first in which I tried this method, and I was rather surprised to see how well it answered. Since then I have treated a good many in the same way and have been favorably impressed with the results. I need not say that if a large portion of the bone has become necrosed before any good can be done the dead part must come away or be taken away.

In the course of an otorrhœa, patients occasionally bring with them portions of the ossicles which have been discharged from the ear; very often the malleus without the processus gracilis, but generally the handle is the first portion to become detached.

I have as yet only made a passing allusion to the healing of perforations which have occurred from disease. When they have been the result of accident they sometimes heal very quickly. The converse is true of the other class, at least, so far as regards adults. By far the larger proportion of these never heal at all, and it is not unusual to see in middle aged and old people perforations which date from infancy; and, indeed, when their mode of production, and the circumstances attending them are considered, it is what might be expected. An accumulation of purulent matter secreted from a mucous membrane lining a closed cavity (for at the time when the pus is first formed the tympanic orifice of the Eustachian tube is swollen and closed), makes its escape by a process of ulceration through the tympanic membrane; so long as the mucous membrane continues to secrete purulent matter, which is discharged through the orifice in the membrane, and which, from the conformation of the tympanic cavity, is especially liable to collect in considerable quantity, it is impossible that the tissue lost in the ulcerative process should be reproduced. The circumstances favorable to this event would be, firstly, that the cavity be freed from morbid matter which acts as a source of irritation; and secondly, that the mucous membrane ceased to secrete. In the case of these conditions being fulfilled, whether as a result of treatment (the object of which should always be to induce this), or from the unassisted reparative powers of nature, perforations heal. When once the healing process begins, it proceeds very rapidly, and I have been surprised at times to see patients in whom a discharge which has lasted for years has lately stopped, and the perforation has healed in a few weeks afterwards.

It is most difficult to get an opportunity of watching this process, and so to make accurate observations on the subject. As a rule, when cicatrization has taken place in the case of a perforation of long standing, the patient has been dismissed when the discharge has ceased, and is only seen again, perhaps, casually many months afterwards, when an inspection of the membrane will show what has happened. Even then there is very little to be seen; the membrane can be observed to be entire throughout, and air will not pass through on inflating the tympanum, but it is most difficult to determine the position of the cicatrix.

In cases of tympanic disease without perforation, whether there is a history of discharge at any time or not, I constantly see what I regard as most likely to be scars; but in deciding this point, it must be borne in mind that in such disease the internal layer of the membrane becomes affected, and that in many instances it is very much thickened and opaque, and presents an appearance altogether different from what it does in health; therefore it is necessary to use great caution in speaking decidedly of particular spots as being cicatricial; and on this question a most accurate observer, whose remarks I have quoted before in the course of these lectures, viz. Dr Adam Politzer, says, "we can speak with certainty of cicatrices only when they have been formed under our own eyes after perforation. We can only conjecturally regard them as such when the patient states that there has been a previous discharge from the ear; while in cases in which there is no recollection of an otorrhœa, the diagnosis between cicatrices and circumscribed atrophies is impossible."

In some cases that I have noted when the patient has been seen after an interval of from one to two years, where the membrane was perforated to as much as, or more than, a fourth of its entire extent; in the position previously occupied by the perforation there has been a calcareous deposit, but it is more usual to find a cicatrix thinner than the rest of the membrane. Two out-patients are now under observation with this condition. Another I saw in the case of a woman, aged 29, who came to the Hospital in October, 1871. On the right side fully one third of the membrane was the seat of a chalky deposit, in the anterior section. Eighteen months before she had been under treatment as an out-patient for about two months for a discharge from this ear. The discharge had some months afterwards ceased, and she had since been considerably more deaf (so she said).

If a patient be seen within a few days of the mem-

brane giving way, and the loss of substance has been
very slight, provided that the tympanum is kept carefully free from secretion, the rupture will not unfrequently heal in a few days. If it does not do so
very quickly the probabilities are greatly against it
doing so at all.

On September 29th, 1871, M. V—, a girl, æt. 20,
attended here with a polypus in the right ear, which
was removed on that day. On the 26th October, she
said that three days before, acute pain had come on in
the left ear. This had lasted without intermission for
two days, and was then succeeded by a discharge and
relief from the pain. There was a little discharge from
the ear, and the watch was heard at one inch from the
auricle. She could blow through a perforation which
could be seen in the anterior section of the membrane;
the action of blowing through separated the sides of
the opening, and a good deal of secretion was forced
in this manner through the perforation, and on sus-
pending the blowing the sides flapped back into position
again.

As the tympanum was in this way emptied of the
secretion, the hearing was improved very much. She
was directed to simply syringe the ear with warm
water twice a day, and on November 2nd the perforation had quite healed, and she was not perceptibly
deaf. I could not detect the scar, although I knew
where to look for it. The membrane was throughout
white and thickened, and the handle of the malleus was
only just discernible.

To how great an extent large losses of substance in

the tympanic membrane may, under favorable circumstances, be replaced by new tissue, seems uncertain;
at any rate occasionally to the extent of half the membrane.   The probabilities of healing taking place when
a perforation takes place in children is very much
greater than with grown up people, and such a probability is immensely increased when the subjects are
infants.   We know how very liable young children are
to purulent catarrh of the middle ear, how constantly
they are the subjects of a discharge from the
ear, which ceases after a few weeks without any
treatment, and how very often an examination proves
that the membrane is entire.   When opportunities of
seeing these little patients arise I have, times out of
number, observed a perforation, and some months or
years afterwards have found an entire membrane.
Provided that the ossicles maintain their position, and
the part of the membrane to which the malleus is
attached remains intact, the plane of the membrane
will not be altered ; but if neither of these conditions
prevail, the support afforded to the membrane will be
in part lost, and there will be more or less collapse.
In this case, adhesions may form between the walls of
the tympanum and the membrane.   From the advanced
position of the promontory this is the part to which
such adhesions are attached. This can be seen
without difficulty when only a part of the membrane
is involved, and the remainder is unattached or absent ;
but when there is what may be called a general
adhesion, where the loss of substance has been central,
the precise state is not so readily detected ; for then,

although a part of the membrane is in truth absent, the loss is now replaced by the promontory. Under such circumstances the deafness is extreme, and no treatment is likely to be of any service.

Having considered so far the subject of perforations of the tympanic membrane arising from disease, let me now say a few words about injuries to the membrane. In the accident ward, with patients who have fallen upon, or received severe blows on, the head, bleeding from the ears is, you are aware, a common symptom, and this may occur whether there be fracture of the base of the skull, or the case be one only of concussion. In both instances the membrana tympani has been ruptured. If the patient live the rupture generally heals in a short time, leaving more or less deafness. If an incision be made with a sharp knife purposely by the surgeon, the wound quickly closes up, sometimes in two or three days. On the same principle, if the membrane be accidentally pierced with a sharp instrument, healing rapidly takes place.

Several cases illustrating this have come under notice here, and I have seen a great many others elsewhere. In one which occurred in 1870, the accident happened to a man of thirty-six. While he was picking his ear with a pin his hand slipped, and the head of the pin went through the membrane. When he was seen, four days after the accident, he was still suffering pain, and had a good deal of tinnitus. There was a perforation at the upper and posterior part of the membrane; he could hear a watch at four inches from the ear, but was very

deaf to the voice. The membrane healed six weeks
after the accident. In another case, when the point
of a pair of scissors was pushed in a man's ear by his
child at play, the portio dura was wounded, and caused
instant facial palsy ; the wound of the membrane
healed very rapidly, leaving a moderate degree of deafness.* The most minute wound of this kind I ever
saw was made in the membrane by a needle, as the
subject of it, a girl of seventeen, was picking her ear.
This healed in a few days, and left the hearing quite
unimpaired.

If a blunt-pointed instrument be used, the deafness
remaining will be very considerable ; inflammation will
be set up in the tympanum, a purulent discharge will
follow, and the same conditions which were observed
when the perforation resulted from disease will interfere with the healing process. In September last a
patient attended here whose right tympanic membrane
had been ruptured by a blade of straw thrust accidentally into the ear. The handle of the malleus was
driven backwards, there was a purulent discharge from
the tympanum, and he was very deaf. He only came
once to the hospital, so I do not know how the case
ended. The handle of the malleus has been fractured
by an accident of this kind, and Mr Toynbee relates
a case where the chorda tympani was similarly
injured.

A loud explosion close to the ear will sometimes
rupture the membrane, or a box on the ear, but this

* This case was reported at the Clinical Society ; see ' Transactions
of Clinical Society,' " Wound of Portio Dura, causing Facial Palsy."

generally happens when the blow is unexpected. I
have seen this injury repeatedly; once with a boy
who unfortunately was totally deaf on the other side.
While he was at play, another boy came behind him
and gave him a sound box on the good ear (the right),
the rupture extended in a line with the handle of the
malleus. He remained very deaf, so much so as to
require a raised voice close to the ear.* I have known
the membrane to give way during fits of violent
vomiting, or in a fit of whooping coughing, or while
the nose was vigorously blowed; in all instances the
deafness was very considerable, and there was suppuration in the tympanum.

In all the cases I have seen when the tympanic
membrane has been accidently ruptured the extent of
impairment which the hearing has suffered is in a very
great measure proportionate to the degree of violence
used. Obviously, more violence will be required to
rupture the tympanic membrane when the force is
applied from without, if the instrument causing such a
rupture is a blunt-pointed one, than when it is sharply
pointed or edged; and in any case, if the force is
applied from within, as in vomiting or blowing the
nose vigorously, or, again, in the example of a blow on
the head, or a box on the ear, the force employed must
be considerable.

If the impaired hearing in these cases depended
solely upon the rupture of the membrane, when
healing takes place in a few days afterwards without
any inflammation in the tympanum, one would expect

* The case was related in the ' Lancet,' 1870.

very naturally that good hearing would succeed to the continuity of the membrane, but experience shows that it is not so.   The true cause of the permanent deafness is, I believe, dependent upon the shock which the nervous structures (the labyrinth) have received at the time of the accident, and this view is rendered extremely probable when we notice that a similar extent of impaired hearing will follow the same kind of accident (a blow on the head or a box on the ear) when the tympanic membrane has not suffered at all.

The fact of the membrane being able to withstand shocks such as a violent explosion or a box on the ear when either were expected was explained by Mr Toynbee as being due to contraction of the tensor tympani.

The treatment required for accidents of this kind is, in most cases, to leave them alone, and prevent the hurtful meddling of anxious friends who advise all sorts of things to be poured into the meatus, and generally with the result of exciting inflammation of the tympanum.   If there is any bleeding, the ear should be syringed with great gentleness; and if the pain following the accident does not shortly subside, three or four leeches may be placed in front of the tragus, and followed by fomentations.   Should the wound not heal, and a discharge be established, the case will require the same management as a perforation after disease.

The results of accidental rupture of the tympanic membrane may be estimated by observing the notes of these cases.   Out of twenty-two cases, the perforation did not heal in ten; eleven healed, and one was in process of healing when last seen; in six cases the hearing did not suffer at all; in the remaining sixteen it was more or less seriously impaired.

LECTURE VII

WE have not yet quite exhausted the subject of
perforations of the tympanic membrane and the many
ills attendant on this lesion. The position of the
portio dura in its course through the tympanum
renders it especially liable to become affected when
this cavity is the seat of inflammation, so that we not
unfrequently find facial paralysis associated with a
purulent discharge from the ear. The aspect presented
by a patient suffering from facial palsy must be
very familiar to all. The paralysed condition of the
muscles supplied by the facial nerve and the conse-
quent effect of the muscles on the other side acting in
antagonism to them combine to give a most distinctive
appearance to the subjects of this malady. There are
three positions in which this nerve may be affected :—
1st. When it is within the cranium ; 2nd, in the
temporal bone ; or, 3rd, superficially after its exit from
that bone. The last of these three forms of the
paralysis is generally attributable to cold, and gets well
sometimes in a few days, or at others it may last for
many months. But it is the palsy that is caused by an
affection of the nerve, while it is in the aqueduct of
Fallopius, that we have just now to deal with. You
will find in the article on this subject in ' Reynolds'

System of Medicine,' Romberg has stated that whenever the source of facial palsy is in the temporal bone
the uvula of the patient invariably points to the paralysed side, and that there is always unilateral paralysis
of the velum palati. From anatomical considerations
such an effect would be almost anticipated, as the motor
power to these parts is certainly derived from the facial
nerve, either directly or indirectly, through the great
petrosal nerve, which is in close connection with the
portio dura in the aqueduct of Fallopius. But be this
as it may, I have certainly seen cases of facial paralysis
clearly depending upon tympanic disease, and have
failed to detect any unusual deviation in the uvula.
Where the disorganisation of the tympanum has gone
to the extent of caries of the bone (this complication
is generally in cases where scarlet fever has been
the origin of all ear trouble) it is not likely that
recovery will take place, and, so far as I know, no case
where this has happened is on record ; but when there
has been profuse suppuration without any evidence of
caries I have noticed on several occasions considerable
improvement when the condition of the tympanum has
become more healthy, still I have never seen so
complete recovery when there has been profuse suppuration that with accurate observation a difference of
the two sides of the face might not be noticed. The
cases to which I wish particularly to direct your
attention are somewhat different to these ; they are full
of interest, and I am convinced that the cause of the
paralysis is oftentimes overlooked. The reason why
it is overlooked is that there is no discharge from the

ear, and, sometimes, only very little deafness to call
attention to the ear. The history of these cases is
brief; an occasional earache with slightly impaired
hearing, followed after a few attacks with facial palsy.
Perhaps the patient is not seen for many months'
afterwards, the pains in the ear have been forgotten,
and the slight one-sided deafness is not mentioned.

Now if the narration of these two symptoms is not
elicited from the patient, the conclusion which the
surgeon very naturally arrives at is that the facial palsy
depends on a lesion of the nerve after its exit from the
stylo-mastoid foramen, and in answer to inquiries as
to the prospects of recovery, gives a favorable opinion,
tells the patient he may hope to be well within perhaps
twelve months (the superficially induced local palsy
seldom lasts longer than this), and the fallacy of his
opinion is proved by time; for it is not a very unusual
circumstance in the cases I am speaking about for the
paralysis to be permanent. I will relate an example of
this in which facial palsy happened some time ago to a
friend of mine, and which is a very fair instance of the
manner in which the affection shows itself and progresses. I saw this gentleman in February last, and
he told me that on the previous Christmas Day he was
in good health and heard well. For two days afterwards he had a slight pain in the left ear, was somewhat deaf, and on the second day he became paralysed
on that side of the face. He had at intervals for a
fortnight afterwards a deep-seated pain in the ear, and
then it passed away. Occasionally, since, he had slight
earache. The hearing was not much impaired on that

side (the left) ; the tympanic membrane was opaque.
I have brought forward this case, not only as an example of the disorder we are considering ; but for an
additional reason. After what I have said in a former
lecture, I feel bound to tell you that this gentleman is
now nearly well, and that he attributes his recovery to
a long course of blistering over the mastoid process,
the treatment being pursued, I need hardly say,
not according to my advice. It seemed to me to be
so very likely that although he was recovering during
the process of blistering, the treatment had so little
to do with it that I was anxious to hear what he
said. He told me that after each blister had risen he
distinctly felt he had more power over the paralysed
muscles. He is a clever surgeon and a man not likely
to be run away with by his imagination. This is one of
the few cases in which I have been able at all to satisfy
myself as to the therapeutical value of blistering behind
the ear.

If you believe that under the influence of mercury
any of the products of inflammation surrounding the
nerve are likely to become absorbed, it will be right
for you to give it, or iodide of potassium with similar
expectations.

The next subject in connection with perforations of
the tympanic membrane is polypus of the ear. When
a discharge from the ear has lasted for a long time,
upon examination may sometimes be seen peering as
it were through the perforation an extremely small,
fleshy, globular tumour, or a somewhat larger but
otherwise similar growth may be in the position usually

occupied by the tympanic membrane; or, again, it may
be so large as to fill completely the external meatus,
and project as an exuberant mass from the outer orifice
of the ear.   In any of these cases, and in the intermediate ones the growth is spoken of as a polypus of
the ear.   It is quite the exception to meet with
these little tumours, unless there is a perforation of
the membrane; indeed, it is to this condition, or more
correctly speaking to the unhealthy state of the tympanum which is present in these cases that polypi almost
always owe their existence.   Cases, however, have been
placed on record where small polypi have been discovered after death in the cavity of the tympanum, the
tympanic membrane being at the time entire.

If the polypus is a large one, and projects from the
ear, it is either covered with epidermis or has a rasp-
berry-like appearance that is very characteristic.

In looking down a speculum where the membrane
has been completely destroyed it is not always easy to
decide at first sight whether the red mass at the bottom
of the meatus is a small polypus or simply granulations
on the lining membrane of the tympanum.   This, however, is ascertained by examination with a probe, when
a polypus will be found to move under the touch.

In structure aural polypi are fibro-cellular and
in proportion to the age of the growth so will the
fibrous element predominate; at least, this is the rule
so far as my observations extend.

A section which has been made from a polypus that
has been hardened with chromic acid will present very
much the appearance of fibrous tissue in the course of

development, and this seems to be true in the cases I have examined, whether the polypus arise from the lining membrane of the tympanum, the tympanic membrane itself, or the external auditory meatus.

Dr Whipham has been kind enough to examine a great number of the polypi which, from time to time, I have removed, and I cannot do better than first quote a description of two he gave me in 1871.

No. 1.—A polypus growing from the roof of the tympanic cavity, projecting from the meatus, and of twelve months' growth. "A fibro-cellular growth, resembling the early form of fibrous tissue. Delicate fibrillated stroma, the fibrillæ having a tendency to arrange themselves in parallel lines. In some parts the stroma is finely granular; cells not very numerous— some round, some oval in shape—beginning to become elongated. Here and there in the section were scattered oil globules."

No. 2.—A small polypus of recent origin arising from the meatus, probably of two or three months' growth. "Numerous small nuclei, scattered without much definite arrangement in a fibrous stroma, of scarcely so delicate structure as in the former case; here and there cells of considerable size, many distinctly elongated where the fibrous tissue is more advanced."

The two accompanying drawings show the structure of polypi of the ear most commonly met with.

Fig. 15 represents a section of a polypus from the tympanum in a woman, thirty years old, removed five months after the attack of tympanitis, which gave rise to the perforation of the membrane. The growth, then,

10

Fig. 15.

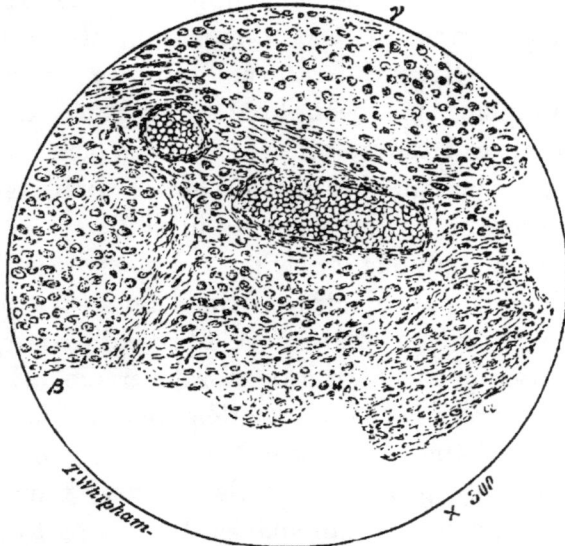

α. Elongated and oat-shaped cells; stroma distinctly fibrous.
β. Round or oval shells, in a delicate fibrillated stroma.
γ. Round-cells in a granular homogeneous stroma, in which slight traces of fibrillation appear occasionally.
    Two vessels are represented in the centre of the field.

must have been of very recent origin, and the cellular character of it is well shown in the drawing.

Fig. 16 is taken from a section of a polypus evidently of greater age. The subject of this is a man, æt. 35, had had a discharge from the ear for fifteen years before I saw him, so that it is impossible to say at what time the polypus began to grow. Except in regard to age, however, it possesses very much the same characters as the one in Fig. 15.

"The microscopic appearances of this growth are

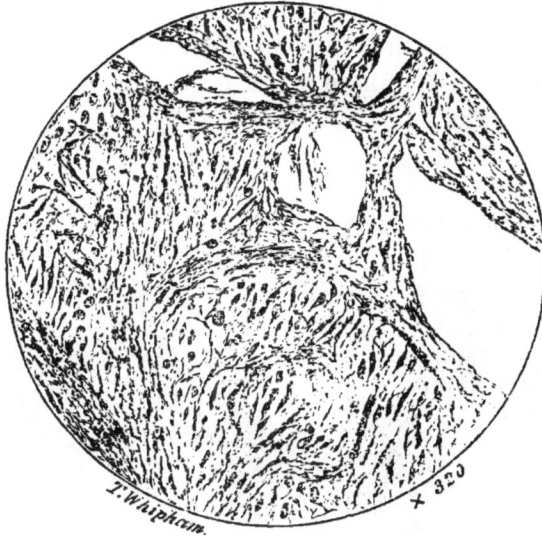

briefly as follows :—The surface of the polypus is composed of round cells, imbedded in a delicate fibrous
network, but without any tendency to linear arrangement. At some little distance from the surface the
cells become elongated or oval, and are placed in rows
more or less parallel to one another. In the central
parts of the growth the cells are considerably elongated
or oat-shaped, and arranged in parallel lines. Perfectly
formed fibrous tissue constitutes the chief portion of
the centre of the polypus. These last appearances are
represented in the sketch."—T. WHIPHAM.

Since this was written, similar structure has been
found in all examined, except in the two following
instances. The first case, Fig. 17 A and B, was a very
large polypus which I removed from the ear of a girl in

December, last year. The growth had been noticed
for many years, and projected for some distance from
the meatus, and arose from the lining membrane on
the roof of the tympanum. I need hardly say that the
membrane was gone.

FIG. 17.

A.—Section from the central part of
the growth.

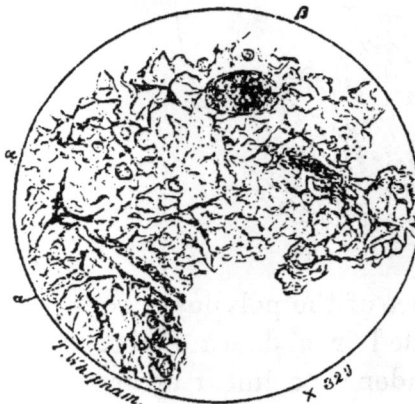

B.

a. Stellate cells.
β. Exudation of serum.

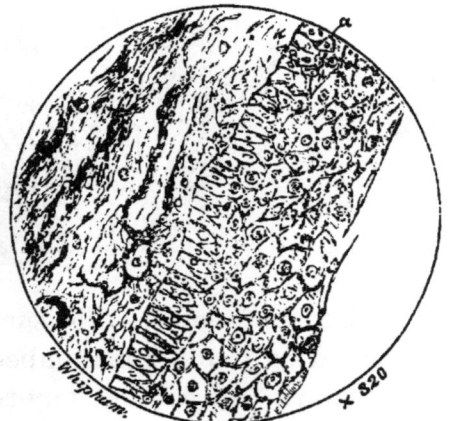

a. Layer of epithelium, limiting the
growth.

"This was an elongated, somewhat lobulated tumour,
of a gelatinous semi-transparent appearance, and very
soft. It hardened rapidly in a solution of chromic
acid.

"Sections examined under the microscope showed
the tumour to consist of a fibrillated interlacing stroma, ·
which in some places was extremely delicate, while in
others it was of a coarser texture. In the meshes

were found, here and there, round cells, thinly scattered; and in other parts the branching anastomosing cells, characteristic of myxoma. Occasionally fibres of yellow elastic tissue were present in considerable numbers, so that the growth answered in great measure to the description of myxoma containing elastic fibres, as given by Cornil and Ranvier at p. 146 of their 'Manuel d'Histologie Pathologique.' However, these elastic fibres in many parts were absent, and then the tumour presented the ordinary appearance of pure myxoma.

" Some qualification of this description is necessary, for although the above mentioned are the chief characteristics of the growth, the stroma presented in some places a rather different appearance to that usually seen in myxoma; that is to say, the reticulum was finer, and the meshes smaller, and the stroma resembled that seen in carefully pencilled sections of lymphadenoma. The absence of lymph cells peculiar to lymphadenoma must, however, distinguish this tumour from growths of that nature. Bounding the tumour was a layer of epithelial cells, which formed in the sections a capsule for the growth. Externally, that is, on the side farthest from the growth, the cells were ordinary, flat, epithelial cells; but on the side in contact with the tumour they were cylindrical, but, as far as could be made out, devoid of cilia. In other parts of the growth fibrous tissue predominated, to the exclusion of the myxomatous characteristics. Occasionally an homogeneous mass (which readily absorbed the carmine) was seen, and was probably due to an exudation of serum from the numerous vessels into the meshes of the tumour. The walls of the vessels were thin and delicate, and the striæ of the arteries clear and distinct."—T. WHIPHAM.

Fig. 18 A and B, represent sections of a growth removed from the tympanum in a girl, æt. 22 ; she had had a discharge from the right ear from childhood, and succeeding to scarlet fever. During the past eight years she had a polypus removed on more than twelve occasions. It would seem from this that it was of a decidedly recurring type. I took away the polypus in September, applied the acid every day for three weeks, and I heard in the following March that there were no signs of regrowth. It remains yet to be seen whether it may be reproduced : so the case is at present incomplete, as showing the results of treatment, but interesting so far as it exhibits a variety of growth different from that usually met with.

"This tumour was composed of an abundant growth of small cells imbedded in a reticulated stroma of extreme delicacy; so closely were the cells packed that it was only by careful pencilling of the section that this stroma was rendered visible. The cells were for the most part round, but occasionally among these one or two of oval shape were seen, no larger than the round ones. In one or two sections, however, tracts were found (running as it were into the proper structure of the tumour) in which the stroma was in some degree fibrillated, and enclosed distinctly nucleated cells, all of which were oval in shape, and at least three times the size of the round cells already de-

scribed. These oval-celled tracts were, however, but
rarely met with.

"These general appearances presented by tumour
are sketched in Fig. A.

"Fig. B represents the appearances seen in certain
parts of the growth. Tubes are seen lined with epithelium, and surrounded by the cell-growth consti-
tuting the bulk of the tumour. The epithelial cells
lining the tubes are more or less oval in shape, and
rather larger than those peculiar to the growth. Their
walls are fibrous, and in some cases very delicate, while
in others they are thick and dense. These tubes have
very much the appearance of gland ducts. They are
represented in the sketch in transverse section, and in
many of them the epithelial lining has been partially
removed during preparation of the section. In the
neighbourhood of these ducts delicate bands of fibrous
tissue are seen running in various directions. The
blood-vessels were few in number, and separated from
one another by wide intervals. Nothing abnormal was
noted in their walls.

"The tumour, then, appears to be a specimen of the
round-celled sarcoma, into the composition of which
the oval-celled variety enters to a very slight extent.
It has more malignant characters than any I have
examined from the tympanum."—T. WHIPHAM.

My experience would lead me to the conclusion that
the lining membrane of the tympanum is the most
frequent situation from which polypi spring; next in
frequency the meatus, and then the tympanic membrane. Whatever their size, structure, or from what-

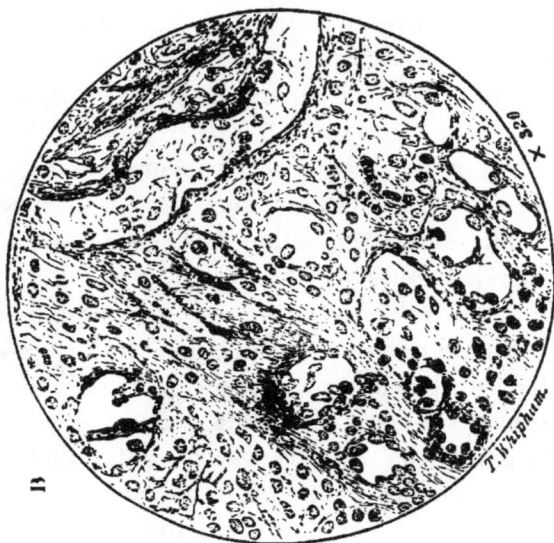

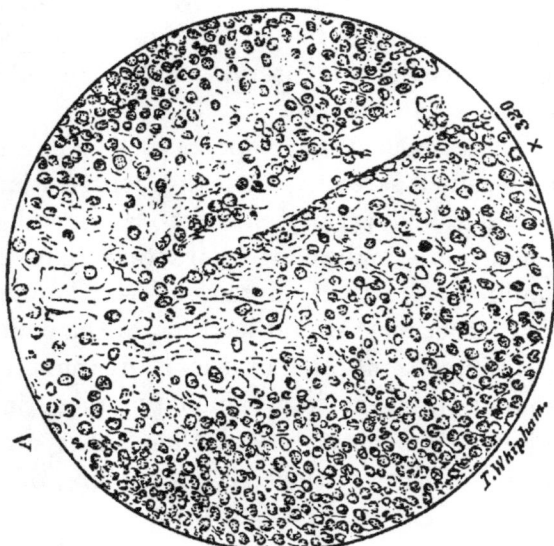

Fig. 18.

ever situation they arise the sooner they are removed
the better.   In the first place, they intercept the
passage of sound to the labyrinth ; in the second, they
keep up a most unhealthy condition of the whole of
the middle ear (there are thus two very efficient
causes for the deafness) ; and, in the third place, they
may become indirectly the cause of death by preventing the free egress of discharge from the tympanum,
thus inducing purulent absorption, meningitis, or
abscess in the brain.

In undertaking the treatment of a case of polypus
of the ear, it is well at once to recognise its possible
tediousness, and the necessary perseverance that may
be required both on the part of the surgeon and patient
to effect its complete eradication.   As a rule it is
perfectly useless to remove polypi and to do no more.
No sooner are they taken away than they at once
commence to grow again, and for this reason their
extraction should be regarded merely as the preliminary
step in treatment.

Although, after removal, a few applications of some
caustic will occasionally eradicate them, and this especially when they arise from the external meatus, the
treatment will often extend over several weeks or even
months, the time occupied depending upon the tendency
to reproduction which the growth manifests.

I am very anxious you should appreciate this, for it
is very satisfactory to thoroughly relieve such an obstinate and troublesome affection as a polypus filling up
the tympanum perhaps, and a profuse purulent discharge from the ear; but it is very unsatisfactory,

both to the patient and yourself, to find that a great deal of time and trouble has been expended, and perhaps pain endured all to no purpose; but if we are to judge from the cases which present themselves where polypi have been taken away from the ear on previous occasions, and have recurred only to be taken away again, and again to reappear, it would seem that the labour required to be expended on these cases has not until late years been fully recognised.

I do not wish it to be understood that all cases of polypus of the ear give so very much trouble. Occasionally a small growth of this kind will come away while the ear is being syringed and never grow again, and in the case of a polypus which grows from the meatus unaccompanied by a perforation, it is just as easy to prevent its regrowth as it is difficult when it arises from the tympanic cavity with a perforation.

When the meatus has been syringed, light should be reflected from a mirror worn on the head as in using the laryngoscope; both hands will thus be free. The growth should be carefully examined with a probe to estimate, as nearly as possible, the exact spot from which it springs. If it be a large polypus a speculum will not be necessary, and a convenient instrument to use is a Wilde's snare. The noose should be made of fine fishing gimp, as this runs more easily through the rings than wire. When it is placed round the growth, and the point of the snare pressed as close to the root as possible, it should be made to cut its way through. If the polypus is not readily seen without a speculum, the largest one that the meatus will hold being placed

FIG. 19.—WILDE'S POLYPUS SNARE.

in it, and the light reflected down it, the growth may
be seized by a pair of rectangular ring forceps and
pulled away. This is not always so easy a matter as it
sounds, at least to get it cleanly away, and especially if
the polypus is a very small one, situated in the cavity
of the tympanum, and perhaps some of the perforate
membrane remaining *in situ*. In such a case it is
always desirable with children that they should have
chloroform, as they will not remain quiet without it
(this is essential), and to ensure complete stillness it
is sometimes necessary in the instance of adults when
the meatus is so narrow as only to hold a small-sized
speculum. In the case of children with small meatus
Toynbee's lever-ring forceps will be found a useful
instrument.

FIG. 20.—THE LEVER-RING FORCEPS OPEN (*Toynbee*).

I wish, however, that it be distinctly understood how
very unimportant it is what sort of instrument is em-

ployed.   As a matter of fact I am in the habit of using almost universally the rectangular ring polypus forceps.   These may be made with blades at various angles, and the rings of several sizes and shapes, and it will be practically found that an examination of the tumour will at once suggest the instrument which will most conveniently grasp the growth.

FIG. 21.—THE LEVER-RING FORCEPS HOLDING A POLYPUS (*Toynbee*).

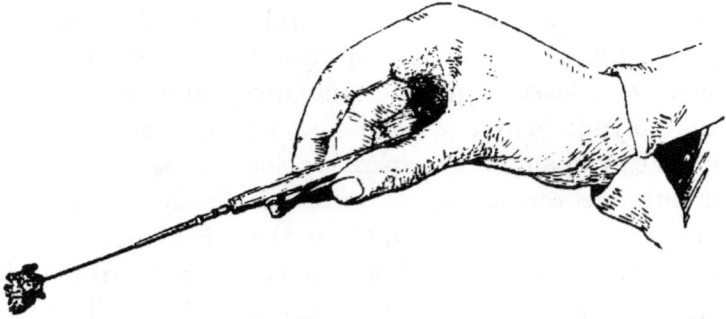

As these little tumours are generally very vascular, there is usually a good deal of bleeding attending their removal, and this is troublesome when the forceps have to be used several times, for it is very often impossible to get all the growth cleanly away at once. It is then necessary to keep on drying the bleeding with pieces of cotton wool so as to get a good view of the part to enclose in the teeth of the forceps.   When every bit of the growth has been taken away from the part from which it sprang, a caustic should be regularly applied until it shows no further signs of regrowth.   The caustic must be a strong one.   Nitrate of silver is too weak.   Potassa fusa and nitric acid are somewhat unmanageable, as they are apt to spread on to parts which it is not desirable should be touched. I generally use chloro-acetic acid, and apply it with a small probe just tipped with cotton wool. Care must be taken not to touch any part of the meatus. If the pain after using the caustic is severe it subsides immediately after using a syringeful of water. Eight or ten applications may be enough, but there can be no rule laid down for this. It is well to use it every day for a few times, and then less frequently. The general treatment will be the same as for an ordinary case of perforation of the tympanic membrane, and should commence immediately after the polypus has been removed. Such treatment you already know is not confined to the meatus, but includes the entire middle ear, except when the Eustachian tube is occluded. Speaking generally, however, I should like to be understood to imply that no pains should be spared to keep the cavity of the tympanum thoroughly cleansed ; and, in case of an adult, no trouble which has this in view should be grudged by the patient.

I believe that the chief causes of failure in the treatment of polypus of the ear may be found, firstly, in the insufficient use of caustics—insufficient, because not continued for long enough periods ; secondly, in the neglect of assiduous care in cleansing the middle ear and inducing a healthy condition of its lining membrane in the ways described in a former lecture. I am led to dwell on the perseverance required in these cases, so that work may not be half done, and that patients may not be dismissed only to reappear some months later with another growth in the same situation as one that has been removed.

Observation teaches me to be more hopeful as to the results of treatment in cases of perforation when they are accompanied by polypi than when they are not. I cannot in any way account for this, any more than I am able to do—judging simply by appearances as regards size, shape, &c.—for the immense variations in improvements to hearing that are met with after treatment in all cases of perforation. When the lining membrane of the tympana and Eustachian tubes becomes more healthy, the improvements that take place with regard to the hearing power are with some very great indeed, while with others they are scarcely perceptible. With these latter a careful examination with the tuning-fork, the history, and some subjective symptoms—tinnitus, worse hearing after fatigue, &c. —will frequently detect a nervous lesion accompanying the tympanic disease.

The following short abstract of seven cases from my note-book will give a very fair idea of the progress and termination of these cases. I have chosen the first five because they show how tedious a process the complete eradication of a polypus in the ear may prove; and the last two, six and seven, as representing the other extreme. In none of them was it found necessary to use either form of artificial membrane.

CASE 1.—June 22nd, 1870. G. P—, a boy, æt. 16. Discharge from right ear from infancy. Watch heard at three inches from the ear, and very deaf to conver-

sation. The natural position of the membrane occupied
by a polypus. This was removed with forceps, and he
attended three times a week till July 25th, and three
times between then and August 14th, when he was
dismissed, the discharge having ceased, and the hearing for conversation being very good. The membrane
on the other side was perforated; but, although the
discharge from the ear ceased, the hearing improved
very little from treatment.

CASE 2.—July 20th, 1870. J. G—, æt. 22. Discharge from right ear after measles in childhood.
Watch heard at two inches from the ear. A fleshy-
looking polypus filled up the meatus. It was removed,
and he was under treatment till August 17th, when a
second small polypus appeared within the tympanum.
This was taken away, and he attended once a week till
Sept. 29th, the discharge having ceased, and the watch
being heard at eight feet, and conversation very well.
In this case, too, the membrane on the other side was
perforated, but the hearing did not improve under
treatment.

CASE 3.—April 19th, 1870. C. F. P—, male, æt. 18.
Discharge from right ear from a child. A large polypus
projecting from the external meatus. Watch not heard
on contact. He attended, after the polypus was removed, three times a week till May 10th; after then
sometimes once and at others twice a week till June
27th; from that date, occasionally at intervals of five
or six weeks for six months. The perforation was very

small, and the cavity of the tympanum became filled with discharge and made him deaf for the time; this caused the case to be so tedious. When dismissed he could hear conversation very well.

CASE 4.—M. M—, female, æt. 22. After scarlet fever. A perforation, and a polypus filling the meatus on the right side. Watch not heard on contact. Polypus removed. She attended once a week for fourteen weeks, when she was dismissed, the discharge having ceased. Although she heard the watch at five feet, the improvement for conversation was not proportionately increased.

CASE 5.—W. H—, male, æt. 19. A perforation and polypus after five years of discharge from the right ear. Watch heard on close contact. Attended twice a week for two months, and was dismissed, the discharge having ceased, hearing increased to six feet with the watch, and conversation very well.

CASE 6.—Where Mr T—, æt. 26, who could hear well four weeks before I saw him, could then only hear a watch at one inch from the ear, and had a polypus growing from the meatus and blocking up the orifice. The polypus was removed, and showed no disposition to grow again, the temporary deafness being only due to closure of the meatus. In ten days no traces of it were to be seen. The membrane was quite healthy.

CASE 7.—E. L—, æt. 26. After catarrh of the

tympanum without a perforation of the membrane. In this example a small polypus could be seen in close apposition to the membrane, and arising from the meatus. This was taken away. Caustic applied five times, and the growth was quite eradicated. The hearing, however, did not improve, the deafness depending on the condition of the cavity of the tympanum remaining after the catarrh.

11

LECTURE VIII

To-DAY we shall consider the modes in which suppuration in the tympanic cavity may terminate fatally. Up to the present our attention has been directed only to the impaired hearing and discharge from the ear which accompanies these cases. Considering the anatomical relations of the tympanum and mastoid cells with reference to the interior of the cranium it need not be a matter of surprise that occasionally suppuration in the cavity of the tympanum should excite inflammation in the cavity contiguous to it. Such a result is sufficiently common to form a few additions each year to the pages of our ' Postmortem Book.' The plate of bone which separates the tympanum from the dura mater is never very thick, and sometimes extremely thin. The mastoid cells have venous communication with the portion of bone which forms the sulcus in which is the lateral sinus, so that if any surprise is excited by fatal contingencies when there is a perforation of the tympanic membrane it should be *not* that they occasionally occur, but that they do not do so more frequently.

In fatal examples of this kind it is more usual than not to find after death that the temporal bone is

carious.   If I relate a few cases taken from the ' Post-
mortem Book,' some of which I had an opportunity of
seeing during life as well as afterwards, and some only
at the examination after death, I shall, perhaps, be
able to give a better general idea of the symptoms
which ought to direct your attention to the ear during
life, and the morbid conditions which cause death, than
by giving a general description of either, for the symptoms vary a good deal in individual cases; at one time
the membranes of the brain are affected, and the
substance at another.

CASE 1.—H. M—, æt. 45, was admitted during the
night of May 2nd under the late Dr Page.   He was
furiously delirious and died the next morning.

Post-mortem examination.—Lymph all over the
surface of the brain beneath the arachnoid.   Effusion
of lymph at the base of the brain, especially thick over
the optic commissures.   The dura mater covering the
upper surface of the right petrous bone was separated
from the adjacent osseous tissue for a considerable
extent, and two small openings were found in the bone
communicating with the cavity of the tympanum.   The
small bones of the ear were destroyed, and the sur--
rounding parts of the petrous bone softened.   The
anterior lobes of the brain were softened, and the
vessels of the posterior lobe gorged with blood.

The only comment I would make about this case is
that whenever a patient is obviously suffering from
head symptoms in the absence of anything else to fully
account for them, attention should be directed to the

ears, and inquiries made as to whether there has been any discharge from either of them.

When a portion of the petrous bone is found after death carious, and the dura mater overlying this part raised, with pus between it and the bone, it is sufficiently clear how the membranes of the brain became affected; but meningitis may attack the subjects of perforation of the tympanic membrane, and at the post-mortem no caries of the bone can be found. This was the case with C. G—, æt. 34, a cabman, who was admitted at 5.30 p.m., on the 23rd of January, 1870, under Dr Barclay. He had complained of pain in the head for two days before. On admission he was sensible; the same evening convulsions came on and he died the next day.

Post-mortem.—Posterior surface of the petrous bone on the left side was dark and discoloured, and on cutting into it the mucous membrane of the tympanum and also of the external auditory meatus was found to be highly congested. There was pus in the meatus, tympanum, and mastoid cells. On the under surface of the brain was a layer of purulent lymph, which was thickest on the left middle lobe and on that part of it which overlay the discoloured bone. What is the medium of communication between the tympanum and the brain when there is no caries of the temporal bone? The same question may be asked when, instead of meningitis an abscess in the cerebrum or cerebellum is found, healthy brain-tissue separating the wall of the abscess from the membranes and temporal bone.

So common a cause of cerebral abscess is suppuration in the tympanum that in Sir W. Gull's and Dr Sutton's article in ' Reynolds' System of Medicine ' out of 76 cases there recorded no less than 28 could be traced to this origin, and in several of these cases the bone was not carious.

The communication must take place either through blood-vessels or the morbid process be propagated through a series of cells. The generally accepted explanation is that the veins are the channel by which the disease spreads from the tympanum to the brain, and if this course were in the direction of the circulation it would be readily understood. The question can hardly be said to be satisfactorily answered by pathologists, but instead of discussing this at present it will be more useful if observation teaches us to be alive to the fact that these fatal events do occur, to recognise the symptoms when we see them during life, and, as far as possible, to guard against such a termination in cases of perforation of the tympanic membrane.

Another fatal accident (so to speak) of tympanic disease is thrombus in the lateral sinus and internal jugular vein, giving rise to pyæmic deposits in various parts of the body. The account which Dr Dickenson wrote in the post-mortem in the following case shows in the most accurate manner how this may take place. The preparation is in the museum.

On December 17th, 1863, a young man, æt. 20, was admitted into the hospital under Dr Fuller with a history of discharge from his right ear dating four months back, and abscess over the mastoid process. He had suffered a good deal at times from pain in the head, and a week before his admission had an epileptiform fit, and this had since recurred four times. He remained in hospital eight days, had a succession of fits, and died on 25th, eight days after admission. At the post-mortem the arachnoid cavity was found full of pus. The lateral sinus of the right side throughout was distended with a light-coloured plug, which, under the microscope, displayed only the materials of broken-down coagulum. The wall in contact with the temporal bone had given way in one place, so as to let out some of the contents of the sinus, and opposite this extravasation and hole was a foramen in the bone situated in the groove for the lateral sinus, which conveyed a vein full of white matter into that vessel. A fine probe could be passed into this foramen, and on cutting away a part of the bone the end was seen in the tympanum, which contained sanious fluid.

Dr Dickenson in writing out the case adds, "It was therefore inferred that the matter had collected in the middle ear, had been taken up by the vein described, and conveyed into the lateral sinus where it had excited suppuration and finally inflammation of the dura mater, deposition of new bone and arachnitis. (The superficial deposit of bone alluded to was in the course of the longitudinal sinus.)

Mr Toynbee laid it down as a rule, and he quoted case after case in proof of it, that after three years of age, whenever the disease was situated in the mastoid cells or the posterior part of the petrous bone, the cerebellum was the part affected in cases of abscess of the brain, and that if it was the superior part of the petrous bone (the tympanum) the cerebrum was the situation of the abscess. That disease usually advances in this direction experience has amply verified, but it would seem that this is subject to occasional variation, as shown by the following case.

On September 26th, 1870, R. H—, æt. 21, was admitted under Mr Prescott Hewett, and on the same day I took the following note of the case:—Ten months ago in good health, when he had pain for several days in left ear, followed by discharge and cessation of pain. Since then more or less discharge. Two weeks before admission, according to his account, the discharge ceased, severe pain came on in the ear and corresponding side of the head. For these symptoms he has been confined to the house ever since The pain has increased. He has slept but very little, and then his sleep has been disturbed, moaning, and (his wife says) he has been delirious at times. At present there is a purulent discharge from the left ear. The tympanic membrane is perforated, a polypus protrudes from within the tympanum and fills up the perforation. Air passed through the Eustachian tube in Politzer's method does not pass through the perforation. The portio dura is paralysed. There is no tenderness over *the mastoid process* or in the centre of the *carotid sheath on that side*. Three days ago had a little shivering, none since. Complains of constant and severe pain over the left temporal region. Intellect somewhat confused, but answers rationally. When spoken to moans. Squints, but says he has done so all his life. General feverish symptoms; skin dry; no sweats.

Mr Rouse was on that day attending the case for Mr Prescott Hewett, and after I had taken this note, at his request, I removed the polypus, which was about the size of a pea, and completely filled the tympanum; for there could be no doubt but that the polypus was and had been preventing the escape of discharge from the tympanum into the meatus. He became gradually worse and died in a comatose condition on October 15th.

The following is the report of the post-mortem examination at which I was present (it is taken from the post-mortem book):

" A large ragged abscess full of fœtid brown-coloured pus and *débris* occupied the greater part of the left lobe of the cerebellum, and projected on its upper surface. At the anterior surface of the petrous portion of the left temporal bone the dura mater was elevated into a prominence of about the size of a walnut, and the cavity thus formed contained fœtid grumous pus. The subjacent bone, though very white, was not softened. The left lateral sinus was occupied by an old clot, which extended to the commencement of the jugular vein, and as far as, or even into, the torcular herophili. No thrombus of any other sinus.

" A ragged abscess existed in the lower part of the upper lobe of the left lung, and had burst into the pleura, setting up much inflammation, and an effusion of recent lymph into the pulmonary pleura."

Thus, besides the brain mischief, there was the additional and sufficient cause for death, as seen in the pyæmic abscess in the lung.

You may, perhaps, remember that this case formed the base of a clinical lecture delivered by Mr Prescott Hewett, in which he laid especial stress upon the fact that the course taken by the disease was unusual, inasmuch as the cerebellum was the seat of abscess, while in most cases, where the disease advanced from the tympanum, it was in the cerebrum that the abscess was found.

Another mode in which the brain may become affected in these cases is by extension from the tympanum into the labyrinth, and thus through the internal auditory meatus into the cranial cavity.

This was the course which the disease took in the case of F. C—, a boy, æt. 12, who was admitted under Dr Fuller on 26th November, 1866. He had had a discharge from the right ear for three years. Ten days before admission he was brought from school with pain in the ear. On the eighth or ninth day after the seizure he had two rigors. When he was admitted he was pallid, with a pulse of 128; temperature 101·5° Fahr., and semi-comatose; a tympanitic belly; rapid breathing, and râles all over both lungs. He was delirious during the night. On the 27th he vomited several times, and died on the 28th.

Post-mortem Examination.—The dura mater round

the internal auditory meatus thickened and congested;
between its layers an abscess full of green pus about
the size of a nut, a superficial patch of pus on the
cerebellum, and corresponding to this spot the petrous
bone carious and softened. Tympanum, internal
auditory meatus, and external ear full of pus.

Both pleuræ contained a large quantity of pus, and
the surface of both lungs was covered with purulent
lymph. Studded throughout were small abscesses,
some the result apparently of softened tubercle, but
many presented the appearance of secondary deposits.

You will have observed in the cases which I have
related that there are several different courses which
disease may take when spreading from the middle ear,
and that the symptoms during the patient's life have
varied considerably. The first premonitory symptom
is very generally a rigor or severe pain in the head, on
the same side as the ear affection. When the brain
shows signs of being involved some patients will
have convulsions, others epileptiform fits, others are
furiously delirious; but the few that I have seen lay in
a semi-comatose condition, moaning and putting their
hands to the head, until by-and-bye they became quite
comatose, and died. From the nature of the disease
its course, you may suppose, is rapid, a few days to a
fortnight generally being about its duration; but
sometimes these cases linger for much longer; one
mentioned in the article I have referred to (Sir W.
Gull's) lived for fifty-three days.

In taking a retrospective and general view of all

the cases I have seen in which suppuration within the tympanum has terminated fatally, I could almost divide them into two classes. First, those in which fatal symptoms come on soon after suppuration has commenced in the tympanic cavity, and, secondly, those in which such symptoms appear after there has been a discharge from the ear for several years. In all, the sudden and unlooked-for ending is sufficiently sad, but when an apparently healthy man or woman, after some hours of what they consider to be earache, has a purulent discharge from the meatus, followed in a few days, perhaps, by a rigor, and dies in the course of another few days, the event comes upon the friends of the patient in so terrible and unexpected a manner that it is well you should make yourselves as familiar as possible with the pathology of these cases, for it will surely happen that sometimes you will find yourselves called upon to see them.

It seems scarcely necessary for me to say that where inflammation of the contents of the cranium has actually commenced, the issue of the case is certain, so that I have nothing to say about treatment; but before I dismiss this subject I have a few words to say about the second division, in which the discharge has lasted for long periods, and to which I ask your thoughtful attention. Several of the patients who have died in the manner we have been considering have suffered from severe and sometimes excruciating pains in the ear and part of the head close around for many days before what we speak of as brain symptoms have set in. Three times during the past two years I

have seen patients here who have been suffering in
this way. In each of these cases a polypus blocked
up the tympanum, and after the polypus had been
taken away the pain subsided in a few hours.

In a fourth case, where head symptoms had been
present for nine days before I saw the patient, he died
four days after the operation.

Of course no one can say how these first three cases
might have proceeded if nothing had been done, but it
seems not altogether improbable that they might have
ended similarly to the case of R. H—, where I removed
the polypus, but not until inflammation of the brain
had become established. If my view of the subject be
correct, whenever the tympanum or mastoid cells are
the seat of suppuration, any circumstance which will
prevent the escape of pus is liable indirectly to induce
a fatal consequence.

What a polypus will do at one time, at another may
be effected by adhesion of the tympanic membrane to
the promontory; or again, a case has been recorded
by Mr Toynbee, in which suppuration has taken place
in the tympanum and mastoid cells, has not caused
ulceration of the tympanic membrane, or sought an
escape through the tissues over the mastoid process.
In this way there has been no discharge externally to
direct attention to the origin of the head symptoms,
and the case has not been cleared up till the post-
mortem examination.

Being fully alive, then, to the dangers which beset
the spread of inflammation from the mastoid cells to
the brain, no time should ever be lost in providing an

escape for pus in this situation whenever it is practicable. Pain and tenderness over the mastoid process should always demand immediate attention, and especially when the tympanic membrane is perforated. If in addition there should be redness or pitting upon firm pressure, it may be assumed with great confidence that there is pus in the mastoid cells. Whether this happens in a child or an adult, a vertical free and deep incision through the periosteum should be made with a strong scalpel about one inch from where the ear joins the head, and a poultice placed over the wound. If the incision is not followed by a flow of pus, the point of the scalpel or a probe pressed into the bone will sometimes break down some diseased bone and let out the matter. In cases of this kind it becomes a question whether in the event of the proceeding I describe not being sufficient to open the cells and so give relief more energetic measures should be adopted. Provided that you can satisfy yourself as to the presence of pus within the bone there can be no question as to the pressing necessity of providing an outlet for it. In order to accomplish this a hole must be bored into the cells, either by means of a sharp-pointed gimlet made for the purpose, or a drill.

The thickness through which it is necessary to bore is often considerable, and the hardness of the bone very great. So soon as the cells are entered a few drops of pus will well up through the wound, and whilst the instrument is being used it is advisable that from time to time the opening is explored with a probe. On the day following the operation the patient

will be able to blow through the wound if both
external meatus are firmly closed, and in so doing will
expel the pus which has collected in the cells.   This
proceeding has often been, I am confident, the means
of placing the patient's life in safety, and a detailed
account of it will be found in a paper by me in the
'Medico-Chirurgical Transactions' entitled "Disease
of the Mastoid Bone."   In the same paper there is also
an account of malignant disease which involved the
mastoid process, besides a considerable part of the
temporal bone.   The patient who was the subject of
this affection died of exhaustion, and in a similar case
which came under my observation the malignant
growth (epithelioma), after eroding large portions of
the bones, opened the internal carotid artery, thus de-
stroying life by hæmorrhage.

In both of these cases the malignant disease in the
early stage closely resembled polypus and commenced
in the tympanum.   The patients had suffered for a
considerable time from discharge from the ear.   The
same course of events appears to have been followed in
all other recorded instances of malignant disease of the
ear.   Thus, the exciting cause of the new formation
was a local irritation in the cancerous as it also is in the
simple polypoid growths in this situation.

In the course of an otorrhœa it is not very unusual
for patients to suffer from symptoms of cerebral irritation, lasting for a few days and passing away, such as
giddiness and pain in the head confined to the same
side as the affected ear ; and in dissections which have
been made of persons who have died from other causes,

and have been the subjects of perforation, the dura
mater lying over the roof of the tympanum has been
found thickened.   How near have these been to a fatal
termination of the disease !   When the tympanum, then,
has been the seat of suppuration, and when there is a
perforation of the membrane, common disease though
it be, it possesses points of interest beyond the impaired
hearing.   Of such, it may be said that so long as the
discharge has free egress, and care is taken to attend
to the ear, there will probably be little cause to fear,
but inattention to the discharge may at any time place
the patient in a position of great peril.   For this
reason a discharge from the ear is regarded by some
insurance companies as an element against granting
a policy, or at any rate demanding an increased pre-
mium.   I can only say that, whenever it is not so
regarded, the companies cannot be said to exercise
very great care of their interests.

LECTURE IX

PERHAPS the most important point to decide in examining the case of a patient suffering from deafness is, how far this symptom is dependent on an affection of the nervous or the conducting apparatus of the ear. Apart from the history and such symptoms as are peculiar to diseases of either part, considerable evidence is often afforded in this direction by the effects produced on the patient, as to hearing, by sonorous vibrations conveyed through the cranial bones directly to the labyrinth, without the intervention of the tympanum. Thus, let a vibrating tuning-fork be placed on the top of the head of a person with good hearing ; after it has ceased to be heard in that position, if it be placed at a little distance from the external ear, it will be heard quite plainly, showing that sonorous vibrations make a greater impression on the auditory nerve when they are transmitted through the conducting apparatus than through the cranial bones.   Again, if the tuning-fork be placed on the vertex, and the external auditory meatus on one side be closed, the sound will be heard more intensely on this side than on the other.   This is also true in respect of the voice of the person on whom the experiment is being made, and in both cases is due to the fact that, when the meatus is closed, the waves of sound, in their passage out from the tympanum through the meatus, are reflected again and again, and therefore their effect on the auditory nerves becomes intensified. Suppose the meatus to be closed with cerumen, or the tympanum to be obstructed with morbid products—the result of catarrh—the same effect will follow ; and, in the case of a patient with the auditory nerve unaffected, he will hear the tuning-fork more loudly on the side which is deaf from these causes, as either interfere with the outward passage of sound. A person in whom the functions of one or both auditory nerves are impaired will hear the tuning-fork (on the vertex) less loudly than he should do in the one or in both ears, and in severe cases will not hear it at all. It follows that, if one ear only be deaf, the tuning-fork will be heard better on this side if the disease is in the middle ear, and worse if it be in the labyrinth. In every case the tuning-fork should be made use of ; but, in estimating the importance of this as an aid to diagnosis, it must be observed that instances are met with in which a person with normal hearing power finds some difficulty at first in hearing a tuning-fork placed on the head, and that some patients cannot at once decide on which side they hear it the louder. The first of these cases is so extremely rare, and the second is so soon overcome with a little trouble, that neither materially affects the value of this test. Still, in making use of this help to diagnosis some little patience is demanded, as at first reliable answers cannot always be obtained from patients. They are apt to answer at

12

random what they think ought to be, viz. that the sound is heard less loudly in the deaf ear. This test is no use in the case of young children.

Next to impairment of hearing, the most common symptom of disease of the ear is tinnitus. It is present in a great variety of affections, alike of the external, middle, and internal ear. It is always an important symptom in making a diagnosis, and although in some cases the cause of it is quite clear, it must be confessed that in others it is quite inexplicable. It seems that any condition which produces pressure on the labyrinth or tympanic membrane may give rise to this symptom. A piece of cerumen lying in contact with the membrane is a familiar example of one, and some cases of Eustachian obstruction of the other. In this latter instance the pressure of air on the external surface of the membrane being greater than on the internal (this was shown in a former lecture to be due to a partial absorption of air in the tympanum), the membrane is retracted, in its turn the handle of the malleus is drawn inwards, and the stapes in this way is unduly pressed on the fenestra ovalis. When the cerumen is removed in the one case, and the tympanum inflated in the other, in the immediate disappearance of the tinnitus we recognise cause and effect, and are able to explain the phenomenon.

At the same time it is quite possible, and indeed very frequently happens, that considerable obstruction of the Eustachian tube with a corresponding amount of deafness may exist without any tinnitus at all, so that although this state unquestionably elicits the symptom in some cases, it does not do so in all. Politzer believes that it may depend on shortening of the 'tensor tympani;' this is simply an effect of prolonged obstruction of the kind named.

If, in cases of catarrh of the middle ear, the tinnitus does not disappear after inflation of the tympanum, it is in all probability due either to a partial or complete anchylosis in some part of the chain of ossicles, or else is dependent on causes situated in the labyrinth. Considering that neither of these conditions is likely to be influenced by treatment, it must be regarded as an unfavorable symptom. Experience bears out this view; thus it is important to give to tinnitus in chronic catarrh its proper significance.

A slight noise in the ears, of which the patient is only conscious when everything around is still and what might also be described as furious tinnitus, repsesent the two extremes, between which are endless varieties in intensity. As a rule, when it is dependent on pressure due either to disease of the middle ear, to impacted cerumen, or foreign bodies in the meatus, it is not of that aggravated character which it assumes when its origin must be sought for in the deeper structures. The nature of the noises is described by patients as resembling recognisable sounds of all sorts, but nearly always they are of a disagreeable kind. Among the few exceptions met with during the past year was a middle-aged woman, who spoke of the noise as a low singing of birds. With persons in whom there is no evidence whatever of catarrh, either from the history or from careful examination of the middle ear, in whom the deafness has slowly come on and advanced to a high degree, tinnitus is a very common symptom ; also in those cases of extreme and total deafness which are met with in the subjects of inherited syphilis.   It may exist for a time with people in whom the hearing power is not perceptibly defective, and with these many causes may be sufficient to produce it, such as periods of annoyance and anxiety, mental fatigue, prolonged suck-ling of children, over-work, taking quinine in large doses : each of these will be sometimes sufficient to induce it.   It is obvious that in such cases it is of nervous origin, and it will generally disappear with rest and constitutional treatment.   In one case under notice in 1870, of a lady, thirty years of age, who had from no apparent cause been subject during the previous year to occasional tinnitus, the noises in the ear came on with the greatest regularity every evening at eight o'clock, lasted for three or four hours, and went away.

Of all the cases I have seen of injury of the tympanic membrane I cannot recall to my mind one in which there has not been, at any rate for some time after-wards, a noise in the ear ; and when the violence used in causing the rupture has been considerable, this symptom has been persistent.   The same may be said of a great number of cases where deafness has followed blows on the head or ear, and violent explosions close to the ear.

It seems not altogether improbable that a little blood may with these patients have been extravasated in the labyrinth, thereby producing pressure from within, but with the same result as when pressure proceeds from without.

Politzer attempts to explain it by supposing that a terminal expansion of the auditory nerve is brought out of its position of equilibrium and is so placed in a temporary or permanent condition of irritation. Both hypotheses are, of course, mere conjecture.

The distress experienced from tinnitus of an aggravated character sometimes almost passes endurance, and, unhappily, in these extreme cases it does not admit of relief. Where there have been opportunities of examining such patients after death, where tinnitus has existed for a long time without any disease of the middle ear, the appearances that have been found on dissection of the petrous bone, by Menière, Politzer, Schwartz, and Hinton, are—disease of the semicircular canals, ecchymosis in the vestibule, hyperæmia of the cochlea, general enlargement and fulness of the vessels of the labyrinth. All these changes would point to distinct pressure within the labyrinth. On the other hand, however, a case was related to me in the General Hospital at Vienna in which a man residing at Trieste, had suffered for many years from tinnitus of so distressing a character that his life was rendered perfectly wretched. All the best aural surgeons in Germany had been consulted by him without any benefit. According to a request made in his will that his ears should be examined after death, a most careful dissection was made of the temporal bones. No abnormal appearance of any kind was detected. In the face of this, and considering that opportunities of examination after death of persons who have been known to suffer from tinnitus are not frequent, we certainly are not in a position to give a satisfactory explanation of this symptom except in a certain number of cases.

In conclusion, then, it seems fair to draw the following inferences : that we find certain conditions coexisting with tinnitus, and so frequently that we may be permitted to recognise in them a sufficient cause : that these conditions appear to produce these symptoms by exciting undue pressure on the fenestra ovalis (and so on the labyrinth), inasmuch as upon the removal by appropriate treatment the tinnitus disappears : that we at times meet with similar conditions unaccompanied by tinnitus : that this symptom undoubtedly exists at others without any apparent cause either manifested during life or on examination after death : that in such cases as the latter we must be content at present to admit our ignorance of the cause, consoling ourselves with the reflection that the confession of ignorance is the first step to knowledge.

One of the commonest histories that will be given by patients who suffer from unmistakeable affections of the nervous apparatus of hearing altogether independently of disease of the middle ear is somewhat as follows :

At some period of adult life they gradually lose the hearing of one or both ears, suffering generally more or less from tinnitus. The deafness will increase with age, and after having reached a certain point become stationary. The tuning-fork placed on the head will be heard imperfectly or not at all.

On examination the tympanic membrane will be found perhaps quite healthy, the Eustachian tubes pervious, no history of catarrh, in short, no origin to the deafness can be discovered. The diagnosis is made then in a considerable degree on negative evidence; by this I mean that beyond the history, tests by the tuning-fork, and perhaps a few subjective nervous symptoms, so long as it is certain that the external ear, the tympana, and Eustachian tubes are quite healthy, it is plain that the lesion which produces the symptom of deafness is a nervous one.

Among other symptoms where impaired hearing is either in part or wholly due to a nervous cause is the fact that the subjects of the affection do not hear so well if they are excited or fatigued, and with women the hearing is very commonly worse during the catamenial periods. But on so many different exciting causes, and under such various circumstances is the power of hearing more or less at times suspended and at others lost, that it is only by carefully observing cases as they come under notice we are able by comparison to recognise separate groups of auditory nerve lesions.

The loss of hearing which follows blows or falls upon the head is clearly due to injury of nervous structure and very little if at all dependent on rupture of the tympanic membrane, if that should take place at the time, for the hearing is impaired by the accident whether the membrane is ruptured or not.

In November, 1870, a labourer fell from a ladder twelve or fourteen feet on to his head, was taken up insensible, and remained so for twenty-four hours; no bleeding from the ears. On recovering consciousness he was deaf on the right side, so much that a watch could not be heard when pressed against that ear.

The tympanic membrane was not ruptured, and healthy looking. There was constant tinnitus, and the tuning-fork vibrating on the vertex could not be heard. The hearing in the left ear was good. No change had taken place in the following January. Twelve months before this in January, 1869, I observed a somewhat similar accident to be followed by severe tinnitus, but no appreciable loss of hearing power. This happened to a housemaid who, while sweeping a room, suddenly raised her head and struck it violently against the bedstead.

Allied to deafness caused by blows on the head is the lasting injury to hearing which takes place after violent explosions near the ear. Artillery men can thus very often trace the deafness which is so common with them to a single shot.

The same kind of accident I have known to happen to men when out shooting in the teeth of a high wind.

If it be true that in such instances as these the tensor tympani contracts when an explosion is expected, and then protects at the same time the tympanic membrane and the labyrinth from injury, this provision is at any rate an insufficient one for saving the hearing in the case of boiler-makers who are exposed to constantly repeated shocks, for a large number of these men notoriously become deaf, after a few years'

work. Without a *post mortem* we can but speculate on
the nature of the injury when a sudden explosion or
blow on the head produces deafness, but again it
seems not altogether unlikely, as in other instances
named, that a little blood may be effused in the labyrinth and the small clot from this have the effect upon
hearing. The concussion occurring at the time of an
explosion would be communicated through the tympanum on to the two fenestræ, and when the result
of a blow on the head, it would be through the
bones.

I have noticed that although the deafness following
a violent explosion close to the ear is often very considerable, a far more extreme degree of it is likely to
follow if there have been any failure in hearing
previous to the accident. I saw in 1870 a most un-
fortunate case of this kind in a man of 50 who was
quite blind and had been totally deaf in the right ear
for many years (after scarlet fever). In 1869, twelve
months before he was seen by me, the hearing of the
left side began to fail and he suffered from slight
tinnitus, he was, however, only a very little deaf.
One day a gun was accidentally discharged near to
him in a small room where he was sitting. He became
not totally deaf to sound, but quite deaf to all
conversation. He said that at the time of the
explosion he felt as if a blood-vessel had burst in his
head (on the left side). I could not discover in the
tympanic membrane any scar, although at the distance
of time at which I saw him, some weeks after the
occurrence detailed, if there had been a rupture of

the membrane it was quite likely that it had left no trace.

Some persons lose their hearing during periods of mental excitement and depression. Generally the defect is symmetrical, and the patients can tell the precise time at which they became deaf—at least so far as this, that having gone to bed one night hearing well, they have awakened the next morning deaf. May it not be that the auditory nerve with such people is in a state which may be described as one predisposed to lose its functions, and that this actually takes place on the arrival of the mental shock? The sudden loss of a relative I have known to act in this way. Such a state is one of disease, and therefore we can hardly attribute the deafness to the incidental cause which awakes it into activity. How else can be explained—what is not of uncommon occurrence—deafness coming on during a confinement, or, what is still more frequent, during suckling? The loss of hearing which occurs during typhus, and is sometimes completely recovered from, and at others remains complete after convalescence, is a familiar example of nervous deafness. The tympanic membrane having all the appearance of health in these cases, distinguishes them at once from those others in which the whole tympanum and membrane have suffered destruction. In instances of this latter kind, where pieces of necrosed bone are discharged from the meatus, the marks of the cochlea occasionally can be seen.

In the 7th vol. of the 'Transactions of the Pathological Society,' there is an account of a case under Mr

Shaw in the Middlesex Hospital where nearly the whole of the petrous bone came away after scarlet fever, and I have myself on several occasions taken large pieces away from the meatus with forceps. Of course in such instances the facial as well as the auditory nerve is paralysed. Perfect hearing may be exchanged for total deafness in the course of scarlet fever in a few hours. Thus a girl, æt. 17, was brought here a few weeks ago. On the 15th day of the fever in the evening (Sunday) she could hear quite well, and on Monday morning she was totally deaf on both sides. Both membranes were quite destroyed, and there was a purulent discharge from the ears; very little time, you will observe, to permit of any remedial measures to the ears during the period before rupture of the membranes took place.

Excepting from such causes as the foregoing, it is not usual to meet with persons in whom, in adult life, the loss of hearing has become total, or, indeed, so severe that the patients cannot be made to understand a word spoken loudly and distinctly close to the ear. Occasionally, however, isolated cases present themselves where it is difficult to trace any cause for the deafness, when there is no history of fever or the like to account for it. The following is an example from a note taken in 1870:

J. D—, a man, æt. 55, went to the West Coast of Africa in 1865, at that time being well. He returned in two years totally deaf. As he could neither write nor read it was impossible to ask him questions; but his wife said that upon his return he had given her a detailed account of everything he had done when away, and had not mentioned any illness, and from what she heard from people who were with him he had had none. He had lost his hearing in a few weeks, so he said, *" but not with fever."* The tympanic membranes were quite healthy.

On many occasions I have known partial or total deafness to take place during an attack of mumps, and inasmuch as this happens not unfrequently in the case of young children, mumps must be included amongst the causes which induce deaf-mutism. This happened in a girl, æt. 7, J. W—, who came here on September 29th, 1871. Her mother said that she could hear perfectly well before she had the mumps in the previous April. She could not hear a sound. From the age of the child, and the total deafness, the idea of inherited syphilis occurred to me at once as a probable cause, but it was negatived by the explanation of the mother, who said that from hearing well on one day, she became totally deaf before the next morning.

Some persons who are very deaf to conversation will tell you that when travelling in a carriage or cab, or by the railway, they hear much better than usual. It has been attempted by some to explain that the improved hearing is imaginary, on the ground that people always raise their voices in order to counteract the surrounding noise. This may, in some instances, be a sufficient explanation, but in others it must be remembered that the speaker is only raising his voice sufficiently for good hearing persons; and I have proved, on more than one occasion, that the increased hearing has been real, and not imaginary, by having it tested with a watch in the carriage. The watch certainly does not change its tick, and a patient who could, under ordinary circumstances (in a room), hear it at one or two inches, has been able to hear it in the railway carriage at nearly two feet. Instances have been related where people who were very deaf could hear fairly well while near the noise of a mill or in a blacksmith's shop when hammering was going on.

No satisfactory explanation has been given of this phenomenon. Before we accept any theories that it is dependent on changes in the tympanum or the tympanic membrane, it must be first shown that it does not occur when no traces of disease can be detected in these positions, and there are, without doubt, many cases in which it is a prominent symptom under such circumstances. To judge from this, and from the fact that even where there has been catarrh of the tympanum the patients who thus hear better in a noise seldom receive much benefit from treatment, in all probability this symptom is in some way or other connected with a nervous lesion.

You see how much we are obliged to depend (independently of what we see) upon the history of the case, and the symptoms in the diagnosis of ear disease, when the seat of lesion is situated beyond the tympanum. You may have noticed, in the examination of patients, how often I have elicited the statement that among the other symptoms which are present when

the hearing begins to fail, is the occurrence of attacks
of giddiness. Of these, instance a healthy man, æt. 55,
seen in October, 1869, who became so deaf in three
weeks that he required a shout near to each ear to
understand what was said. During this time he had
attacks of giddiness daily, sometimes twice in the day,
so severely that he was obliged to lie down for an hour
at a time. There was incessant tinnitus. No disease
could be discovered in the middle ear. Another case
was seen this year, in which a woman of fifty years of
age had similar attacks for two weeks, and during the
time became nearly totally deaf in one ear. This
symptom, which will be found recorded in the case-
book over and over again, in the notes about patients
who have no evidence of disease of the tympanum,
does not generally persist after the loss of hearing has
ceased to progress.

In 1861 it was noticed by Menière to be of such
frequent occurrence that he gave the subject of vertigo,
happening in the course of ear affections, his careful
attention, and at the present time, patients who exhibit
a certain train of nervous symptoms are spoken of as
being the subjects of Menière's disease. It is of great
importance that you should be familiar with this affection, as the knowledge may some time or other serve
you in good stead, for otherwise you might suppose a
patient suffering in this way to be the subject of a
brain affection.

One of the most instructive cases of this kind I have
seen occurred in the person of an active, vigorous,
healthy lady, æt. 48 years. She was seized with an

attack of giddiness whilst she was in a shop, and had
to be assisted into another room, where she lay down
for an hour and a half before she was sufficiently recovered to be driven home; this happened in November,
1871. When she got home she was placed in bed,
and so remained for four days. During this time the
only symptom that she complained of was that of extreme giddiness coming on at the least attempt to
move. This gradually subsided, but for three weeks
she could not trust herself to walk alone. On the day
of the seizure she found that she had become very deaf
on the right side, and had some tinnitus in that ear.
I first saw this case in the following January, and in
giving this clear account of herself she said that the
deafness or tinnitus had not perceptibly increased since
the first illness. The middle ear presented no signs of
disease, and a tuning-fork on the head was heard on
the right side, rather less well than on the other. In
the April following, when I again saw this lady, there
had been no change in the hearing, and no further
attacks of vertigo.

A very much milder case, but evidently belonging
to the same class of disorder, happened under my notice
in a gentleman, æt. 53. His hearing was perfectly
good until an attack of giddiness, which came on while
he was at breakfast, and lasted for about five minutes,
leaving him very considerably deaf in the left ear.
He had no further recurrence of the vertigo, and the
deafness remained stationary afterwards. I will give
you one more instance resembling, in some particulars,
these two cases, and then I shall have done.

A woman, æt. 32, totally deaf with both ears, came here with the following account, which was related in part by herself and in part by her sister. Five years and a half ago she suffered from extreme and constant pain in the head and both ears for a period of six months. One morning she awoke, hearing well as usual, and on getting out of bed she was seized with an attack of giddiness; in the course of four or five hours she had lost all hearing power. For one month after this she could not walk without assistance, as the fits of giddiness recurred occasionally; the pains in the ears never troubled her after the attack. The external and middle ear were both healthy. Now these are fair samples of what is called Menière's disease, and the nervous lesion must be situate in the brain, or in the labyrinth.

In the opinion of a great many observers the seat of disease in these cases is the semicircular canals. This supposition is mainly, I may say almost entirely, based upon the results obtained by the well-known experiments of Fleurens in 1842, and subsequently of Goltz in 1870, which have been considered conclusively to show that any injury to the semicircular canals is always followed by loss of power in maintaining the equilibrium; so exact, indeed, were these results stated to be that certain definite kinds of movements followed the injury to the horizontal or vertical canals, or both conjointly. Quite lately, however, the results of all these experiments have been declared to be fallacious by Dr. Böettcher of Dorpat, who attributes the movements of the animals which followed the division of the semicircular canals to injury inflicted on the brain in the immediate vicinity of the canals during the course of the experiments; Dr Böettcher was able to divide one of the canals without affecting the equilibrium of the animal operated upon.

Until, therefore, the functions of this part of the labyrinth have been more unanimously agreed upon by physiologists, it is somewhat premature to speak with much confidence as to which is the exact position of the morbid change in these cases. To judge from the complete disappearance of head symptoms, and the subsequent good health of the patients, it would seem more probable that the labyrinth was affected than the brain. It is only by obtaining a clear account of the symptoms during life, and examining the condition after death, that the question is likely to be settled in a satisfactory manner. Up to the present, owing to the difficulty of getting post-mortem examinations of persons who die, perhaps, many years after the accession of deafness, this has not been done.

The practical lesson to be deduced from these few remarks on this subject is the importance which should be attached to a history of the case, as it will influence an opinion with regard to the chances of improvement taking place in the hearing, or, on the other hand, of the patients getting worse.

So far as I have noticed, when giddiness has been a prominent feature in deafness from nervous causes, where there has been a single severe attack of giddiness followed by impaired hearing, the deafness has been very considerable, and this whether one or both ears are at the time affected. Also, that where less severe attacks of vertigo have succeeded one another, after each seizure in most cases the hearing has suffered diminution, and when these periodical fits of giddiness have discontinued, the hearing has not suffered any further impairment.

Bearing in mind the intimate relation between the auditory and the pneumogastic nerves at their origin ; remembering how the pneumogastric nerve may be excited by irritation of the ear, for example, in syringing the ear in cases of perforation, when the patient will occasionally suddenly vomit without any premonitory feeling of sickness ; how irritation of the pneumogastric nerve will induce unsteadiness of gait with a with a feeling of nausea, I would suggest for your consideration whether in these cases of so-called Menières disease the lesion may not begin at the origin or in the course of the auditory nerve, and the symptoms of giddiness and sickness be due to reflex action excited in the pneumogastric nerve.

Lastly, in connection with these cases I am not acquainted with any treatment which is likely, with any degree of certainty, to ameliorate the deafness or tinnitus. Some of them have improved while the patients were taking small doses of strychnia, but whether it has been due to the medicine or to the influence of time is quite doubtful—I should think the latter.

LECTURE X

In 1863, Mr Hutchinson drew attention to the pre-valence of deafness in patients affected with inherited syphilis.

Out of twenty-one cases which he relates, in the last six there was no indication of disease in the outer or middle ear which could account for the deafness, and the affection was therefore put down to disease of the auditory nerve. In the other fifteen cases the ears were not examined; but in many there was a history of otorrhœa, and to however small an extent the symptom of deafness may have been due to tympanitic disease, there was, at any rate, sufficient evidence of its presence to prevent the cases from being classed as purely nervous.

Of the following cases of deafness in patients the subjects of inherited syphilis, in none was there any evidence of tympanic disease; for I think, in considering the pathology of the nerve affection, it is desirable to have cases from which all other sources of the deafness may be safely excluded. In these twelve, then, there was no history of catarrh, air freely entered the tympana by the Eustachian tubes, and the tympanic membranes were healthy.

CASE 1.—G. C—, a boy, æt. 10.   Keratitis at eight years of age ; could hear well eighteen months ago ; since then been getting deaf ; now can hear a loud and distinct voice if close to the ears; typical teeth; tuning-fork on head not heard at all.

CASE 2.—M. K—, a girl, æt. 12.   Thorough syphilitic physiognomy ; teeth questionable ; she could hear well two years ago, when she had keratitis in both eyes and about the same time began to get deaf.   She was not, however, very much so—that is, she could hear conversation fairly well until within four weeks before I saw her.   During that month she became rapidly worse, and was then totally deaf to any sound.

CASE 3.—J. P—, a boy, æt. 13.   Teeth well-marked ; three years ago keratitis, five months before could hear well ; since then been gradually getting deaf ; requires a loud voice in conversation ; tuning-fork not heard at all through the cranial bones ; watch not heard in contact with ears.

CASE 4. J. T—, a boy, æt. 14.   Quite blind from interstitial keratitis for several years; gradually getting deaf for a year on the left side, but up to two or three months ago could hear well with the right ear.   The hearing now is perceptibly better on this side ; but on both sides requires a raised voice to make him hear ; the tuning-fork could not be heard on the head.   This boy maintained that his hearing improved after taking iodide of potassium for six weeks, and begged to be allowed to continue the medicine, but although he persistently declared he could hear better I could not detect the slightest difference on testing the hearing; typical teeth.

CASE 5.—J. H—, a boy, æt. 12, with well-marked syphilitic physiognomy and teeth. No keratitis; could hear well three years ago; gradually getting deaf; can hear tuning-fork if very loud on the head, and watch in close contact with either ear.

CASE 5.—H. L—, a boy, æt. 10. Syphilitic physiognomy and teeth; mother with scars of rupia on face; she had four miscarriages after she was married, and this boy is the only child. He has been getting deaf for five years; can just hear a loud shout close to ear.

CASE 7.—L. M—, a woman, æt. 26, at ten years of age had keratitis; three years ago could hear well, and in a few weeks became as she now is, totally deaf; suffers very much from tinnitus. The physiognomy of hereditary syphilis well-marked.

CASE 8.—K. W—, æt. 9, a girl. Typical teeth; keratitis five years ago; cornea now affected; slowly getting deaf for four years, and for last twelve months has not heard a sound.

CASE 9.—M. A. W—, æt. 17, a girl with typical teeth; marks of rupial scars on the mother; could hear quite well three week before; now can hear loud clap of hands but cannot distinguish a word; for all practical purposes totally deaf.

CASE 10.—E. H—, æt. 13.   Keratitis at three years of age ; been getting deaf very gradually from that time ; tinnitus constant and most distressing ; tuning-fork not heard at all ; requires a loud shout to make her understand a word ; speaks very indistinctly owing to having been a little deaf at any early age, and the parents probably not taking pains to teach her to articulate properly ; she has rather better hearing on the left side.

CASE 11.—J. G—, a girl, æt. 17.   Three years ago could hear quite well ; at this time she had keratitis, and simultaneously began to lose power of hearing ; she is now totally blind and deaf.   Teeth typical, and physiognomy of inherited syphilis.

CASE 12.—A. H—, boy, æt. 12, at ten years of age began to get deaf, and in eighteen month became stone deaf; typical teeth ; interstitial keratitis. Mother been confined eight times ; first child lived three hours, second lived one hour, third stillborn, fourth lived one hour and a half, fifth is present case, sixth, seventh and eighth, good hearing.

It will be observed that the most general time at which this affection shows itself is from five to fifteen years of age ; the earliest time at which I have seen it being five years, the latest twenty-three.   The course of the disease varies, but in chief part it is rapid in its progress. In the Case 9, the girl, from having perfect hearing, became totally deaf in three weeks. In others it is more slow, taking sometimes four or even five years to arrive at its most severe point. Cases of this kind are sufficiently easy of diagnosis by the history and course of the affection, by the absence of causes in the outer and middle ear for the deafness, by the distinct evidence of disease existing in the nerve as shown by the tuning-fork not being heard through the cranial bones, and by tinnitus.

This morbid condition of the auditory nerve would seem, therefore, to differ in some respects from the others most commonly met with; in the latter kind, with some very few exceptions, the deafness does not come on so early in life, seldom appearing before the age of puberty, neither does it advance so rapidly as in the syphilitic variety, nor, excepting in very rare cases, does it become so complete. It is important to recognise these points, for it frequently happens in examining obscure cases of affection of the auditory nerve that by noticing such circumstances one is led to look for evidence of a syphilitic origin, and in case of discovering it to arrive at a satisfactory conclusion. It may be urged that there will still remain a certain number of cases in which the symptoms throughout are not distinguishable from those under notice, and the following may be quoted as one of these. In September, 1869, I saw a girl, æt. 15, who was totally deaf. The tympanic membranes and Eustachian tubes were healthy. She could hear quite well four years before, and up to two years back could distinguish words if spoken very loudly into the ear.   There were no indications of a syphilitic taint, but the history, the time of life at which the deafness came on, all point to the disease under consideration.   In the absence of the usual signs this case would not admit of being placed among the syphilitic ones; still one is led to conjecture on the possibility of the loss of hearing being subject to the same cause, and the fact of such examples occurring (although very rarely) in children apparently otherwise healthy ought not to deter observers from noting a nervous disease in the subjects of inherited syphilis as one *sui generis*.

In his remarks on this affection Mr. Hutchinson, after observing that the loss of hearing is always symmetrical (with one exception, No. 4, this held good in all the cases I have seen), classes them as analogues of syphilitic retinitis and white atrophy of the optic nerves, and concludes by deprecating any treatment as likely to be useful.   The experience of other observers as well as my own has since shown the correctness of this conclusion, and inherited syphilis as a cause of deafness is now fully recognised.   Indeed, it is the exception for a week to pass by without seeing these cases in the out-patient room.

The subjects of constitutional acquired syphilis, during the course of what are spoken of as secondary symptoms, not unfrequently suffer from more or less deafness which disappears generally under the specific treatment adopted in these cases.   The evidence that the symptom of deafness is dependent on the syphilitic poison is usually conclusive, and is twofold. In the first place the middle ear being found on examination quite healthy, the vibrations of a tuning-fork on the vertex being heard indifferently or not at all, and the loss of hearing being symmetrical, points to a nervous lesion. In the second place, the hearing generally returns as the patients recover from the disease, no treatment having especial reference to the ears being required. As a rule the deafness is not very extreme, that is, the patients can hear if spoken to in a loud voice, and to this rule I have, however, seen exceptions, and notably in the case of a man, æt. 34, who without a trace of disease in the external or middle ear lost his hearing gradually on the appearance of a syphilitic psoriasis, and recovered slowly in the course of six months. At the time when he was most deaf it was almost impossible to make him hear a word. The only local syphilitic affection of the middle ear I have noticed is an occasional obstruction of one or both Eustachian tubes (and this has readily yielded to simple measures, such as Politzer's inflation) at the time that there was secondary ulceration of the throat. It is not very uncommon for a patient with perforation of the tympanic membrane to develop mucous tubercles in the external canal if he becomes syphilised, and this, of course, is due to the fact that the discharge from the ear becomes syphilitic when his constitution is imbued with the disease.

However various may be the symptoms in nervous lesions of the auditory apparatus at any rate the diseases of the middle ear present certain characteristics which are made sufficiently evident by the methods of examination at our command to enable us at once to separate into two distinct classes affection of the conducting and nervous parts of hearing. So at least you might think, and with considerable justice too, after the accounts you have listened to of both kinds; so I sincerely wish was invariably the case. But experience forbids me to overlook the occasional variance from this; very occasional it is true, but on that account none the less necessary for us to be able to recognise as an undoubted fact that the two kinds are sometimes strangely mingled. Such cases are not a little embarrassing, and it is only by constantly and carefully examining case after case as they are brought under notice, that you will be enabled to detect them. An example of what I mean will be such as the following:

A young girl of 18 or 20 after a succession of colds becomes somewhat deaf. The tympanic membranes have lost their translucency and are opaque. On the accession of colds the hearing is worse, some slight and transitory improvement will follow perhaps inflation of the tympana. The hearing will be worse in damp weather. The tuning-fork will be heard perfectly well on the vertex, any tinnitus if present will be temporary and trifling, and not more than is often present with an ordinary tympanic catarrh.

With such a case the usual methods of treatment ought to relieve the patient, for the disease has not been of so long a duration that any remains in the tympanum of old secretion cannot reasonably be expected to be got rid of. Regular inflation and other treatment is perhaps practised for some time, but with no benefit. All this is disappointing to the patient, and scarcely less so to the surgeon. By-and-by you will slowly learn, for it is not always easy to elicit information from nervous girls of this sort, that she is for a day or two during the catamenial period dreadfully deaf. That perhaps on coming home late from a ball after an evening of excitement she can scarcely hear anything that is said to her, and for the following day still remains very deaf. That peculiar aspect and manner which I cannot explain, but which is really symptomatic of some forms of nervous deafness, perhaps is now noticed on some occasion. The patient goes away, and is seen no more for a time.

Now, if the career of this girl (I am putting an imaginary case, but one the like of which I have several times seen) could be followed out, it is in the highest degree probable that, supposing her to get married and to become a mother, she will during her first confinement suffer from an increase of deafness and not improve after convalescence. Each successive confinement will be marked by a fresh accession of deafness, until at last it is very severe. Now, the vibrations of a tuning-fork placed on the vertex are not heard well, and there is a good deal of tinnitus. On the first examination it was not at all easy to recognise the combination of the two kinds of affection, viz. the tympanic and nervous. So far as the history and appearance of the membrane would serve as guides to a diagnosis, there was nothing discoverable byond a tympanic catarrh. I do not mean to say that all these cases progress so sadly as the typical one I have mentioned, but they do sometimes, and therefore they deserve notice.

It would almost seem that with them, as in some others mentioned before, the auditory nerve was in a condition especially predisposed to lose its functions, and that upon any circumstance arising which suspends for a time perfect hearing, such as the tympanic catarrh, it becomes affected permanently. This may appear a lame way of accounting for what I can give no better explanation of, and you will find that another and apparently a more rational one will be given by some others by putting out of the question any nervous lesion, and attributing all the symptoms detailed to a local cause resulting from the catarrh, such as thickening of the lining membrane of the tympanum, and of the fenestræ rotunda and ovalis. I cannot accept this as a true explanation, and for this reason. At the time I speak of, when the patient is first seen, the disease of the tympanum is not only inactive, but all that interferes with the conduction of sound is strictly confined to what has been left behind after the subsidence of the catarrh. The deafness from this cause then ought to remain stationary, and it does so, excepting on the accession of some additional circumstance which does not reawaken the catarrh, but without doing this or in any way interfering with the conduction of sound to the labyrinth induces a decrease in the hearing power.

Although in truth there is not much to be done in

the way of treatment for these patients, and nothing of a local nature, you will warn them to avoid all habits or excesses which are of a nervously exhausting nature, and as apparently slight causes will sometimes plainly have a decided effect on the hearing, so will judicious care sometimes, so far as we can judge, act in a beneficial way in preventing an increase of the deafness. Sometimes I have thought a course of strychnine has acted in this way and on more than one occasion when the hearing has begun to fail under severe mental work I have known good hearing to return after a complete rest of a few weeks.

It is notorious that some families inherit deafness; by this I do not mean that the children are born with defective hearing, but that in early life they gradually become deaf in one or both ears without any discoverable cause for the fact. It has probably been only a coincidence, but I have more often than not observed this to run in the female line, and the most noticeable example of this was (and a very remarkable circumstance it is too) in a family where for three generations nearly all the children on the female side became more or less deaf about 12 or 14 years of age, the only exception being one set of cousins, and these having now reached adult life hear well.

What a confession of ignorance pervades these last two lectures so far as a complete knowledge of the pathology of some of these nervous affections of the ear is concerned! What a confession of the impotence of surgery or medicine for their relief! One thing, however, I hope you have learned, viz. to recognise these conditions when you meet with them, to form a fair notion of whether they are of a progressive or of a stationary nature, and so by explaining what you know to your patients, relieve them from the miserable apprehensions they may have of becoming worse when there is no cause for fear, and by assuring them of the incurable character of their symptoms, save them from the additional annoyances they might suffer at the hands of charlatans who with all the confidence of ignorance are profuse in their promises of what they call a cure.

LECTURE XI

You are doubtless aware that deaf-mutism is dependent on several different conditions; but, speaking generally, it may be said to be caused either by total or partial deafness, the result of congenital defect or of disease occurring during early life. We will take in order the various causes of mutism.

1. Of course, a child who has never heard cannot acquire language in the ordinary way, and therefore is dumb. The chief defects that have been found in the ears in congenitally mute children are a complete absence of the labyrinth; the cochlea with only one turn and a half; absence of the semicircular canals; one or two of the semicircular canals imperfect; the fenestra rotunda closed by bone.

If a completely deaf child be brought to you, you will not always find it very easy to persuade the parents that their child has never heard, or, indeed that it does not hear a little. The point may, however, readily be put at rest in the following way :—While the child's attention is taken up by some toy or the like, and you make a very loud sound, such as a whistle behind it, at the same time taking care that your shadow does not cross its field of vision, or your breath fall upon it, if it does not show some sign of hearing, such as looking round, it may be considered totally deaf. Because the child makes some labial sounds "Mam" and "Pap," its friends do not understand how these words can have been learned without they have been heard by the child. They forget, or rather they don't know, that by watching the lips the two labials M and P are at once without any trouble learned by mutes, and that as these two words have been often repeated to the child the motion of the lips necessary to execute such sounds have been copied by it. Mute children too are singularly susceptible to vibrations, and will turn round when any one enters a room or shuts a door, thus again giving rise to the fallacious idea that they hear, so that in practice it is well to be aware of this little difficulty.

2. It is not at all necessary that there should be a complete absence of hearing power in every case where a child has never spoken; a very moderate degree of deafness is quite enough to produce this misfortune. When it is remembered that children acquire language by hearing each word frequently repeated, and that every new word they learn is imitated imperfectly at first, and gradually the articulation is corrected after successive attempts, it will be readily understood how in the case of a child who was rather deaf, the only way in which language could be acquired in the ordinary way would be by repeating a word over and over again in a loud voice close to the ear which is possessed of partial hearing power, and correcting the articulation of the child until it was perfected. This process not

having been pursued, as a natural consequence the child is dumb.

3. The next example is where a child is born with good hearing power, and loses the hearing either totally or in a great measure before it is old enough to have acquired speech. Scarlet fever and other of the exanthemata are the most common causes of this.

4. A child may have learned to talk, and at the age of 3, 4, 5, 6, or even 7, may become totally or extremely deaf; in the course of a few weeks in the instance of a child of 3 or 4, in a few months if the child is older, it will be dumb.

The most fruitful causes of deaf-mutism induced by disease at this period of life are scarlet fever and inherited syphilis, but I have also known children to become nearly dumb when the Eustachian tubes have been obstructed very extensively and for long periods; thus you observe a moderate degree of deafness sometimes quite suffices to act in the same way as complete absence of hearing power does at others. When children are in process of becoming dumb, their articulation gets more and more indistinct, until by-and-by it is impossible to make out what they say, so that whenever you see a child who is at all deaf beginning to speak thickly it is well to remember the danger which threatens, and without delay to turn your attention to the condition of the ears; for it is surprising how quickly the power of articulate speech becomes lost when it once begins to fail, and the reason of this is sufficiently obvious; the child not hearing what is said around very soon forgets the knowledge that it has not prac-

14

tised itself in exercising for a long enough period to make speech a confirmed habit. The same influences which make a child in this manner dumb will cause a child to lose one language while it is acquiring another. Thus a boy or girl of 4 or 5 years of age who has been brought up in India with a native nurse and taught as a first language Hindostanee, will in six months if it is brought to England and does not hear this language spoken have completely forgotten it.

After examining a case in which the speech has been lost or not acquired in consequence of impaired hearing, the first question to determine is whether any treatment is likely to improve the hearing, and, if so, to what extent it probably will be restored; for in case of a fair extent of hearing being arrived at, no further treatment or management of the child will be required, and in the natural course of events the child will learn to talk. An interesting case is recorded by me in the 'St. George's Hospital Reports,' 1872, in which a child acquired speech in the ordinary way, became considerably deaf after scarlet fever with perforations of the tympanic membranes, rapidly lost all power of speech, and after recovering the hearing in a great measure under treatment learned to talk for the second time in her life, in the course of six weeks. The rapid way in which this child, eight years of age, acquired the power of speech, is a great contrast to the slow and tedious process pursued by the same child when very young, and was plainly due to superior development of her mental capacity. In former lectures the whole subject of treatment of the ear affections has been dwelt upon,

so that for the present we will dismiss from consideration the cases in which the hearing can be benefited,
and briefly discuss the management of dumb children
who are either totally deaf, are partially and incurably
deaf. A question of very great importance to the
parents of mutes is, which is the best system by which
the child or children should be educated.

You know that until very recently, indeed, the plan
adopted in this country has invariably been one by
which children are taught to express themselves by the
use of the finger alphabets (Dactylology) and manual
signs, in common parlance to talk on their fingers.
Mutes who have been taught on this principle, after
instruction extending over a brief period are able to
converse with other mutes and those who know their
language, and when they have acquired facility can
hold a conversation almost as quickly as can speaking
persons. Another system has for many years prevailed
in Germany, Holland, Austria, Sweden, and some
parts of the United States of America. The finger
alphabet and artificial signs by gesture are not made
use of, but the children are taught to use articulate
speech, and by watching the lips of others understand
what is being said. This is achieved by diligent culti-
vation of the powers of observation and imitation. It
is the duty of every medical man to make himself
acquainted (if opportunity offers) as far as possible,
with the results of both systems, so that when he is
asked which he recommends he may be able to give an
opinion on the subject.

With regard to the plan usually pursued in England

I have nothing to say, as the working of it may be seen by paying a visit to any deaf and dumb institution, but of the other, as it is not so familar to persons in this country I have something to say, and I cannot more rapidly explain myself than by reading an extract taken from an account I have elsewhere given of this subject.*

"The so-called German system of education of mutes may be briefly described as one where deaf and dumb children are taught to understand and employ language, by observation and imitation of the articulation of others; the finger alphabet and all artificial signs being rigidly excluded.

"For this, as for any other system, it is of course necessary that the child's intellectual faculties be not more than usually deficient, and obviously where there is a cleft palate or other malformation of the organs of speech (which appears to exist in the proportion of one in one hundred mutes), it is not applicable. The age at which education commences is about seven years, and eight years are expended before the child can read from the lips of ordinary persons, and speak so as to be easily understood by them. Although artificial signs are excluded in the education, it is permitted and indeed necessary at the very commencement, to attract the child's notice by pointing to the teacher's lips, and to various objects, in order to excite and ultimately obtain its undivided attention, as it is from the exercise of this, and from the child's inherent power

* *Vide* 'The Education of the Deaf and Dumb by means of Lip-reading and Articulation.'

of imitation, that all its future education is to be derived.

"To begin at the earliest lesson of a mute of seven years, who has received no sort of instruction. He is brought into a room, when a hearing person is spoken to by the teacher. The child soon notices that as the teacher's lips move, the listener turns round and looks at him, and he thus learns to have his attention directed to the lips of his instructor. Without entering at any length upon the subject of sounds and letters as taught to mutes, it will with a little consideration be seen how, though at first sight it is a difficulty to elicit proper sounds from them by placing their lips and tongue into the necessary positions, it is by no means an insurmountable one, and that a very complete alphabet of sounds may be formed, so that as the pupil progresses with the alphabet, he is taught in a short time by joining two sounds to articulate a word. As soon as this first step is accomplished, then the attention of the child being at once directed to some object or picture which represents the word pronounced, the object after a little time becomes associated in his mind with the sound which he has made to correspond to it.

"By way of illustration—one of the earliest lessons. The mouth of the child being opened, he is made to effect an expiration. This is done, firstly, by his imitating the teacher, and, secondly, by the latter exerting at the same time a little pressure on the epigastrium of the child. Thus, the sound which corresponds in the phonetic alphabet to the letter *h* is evoked; and it is to be noticed, for reasons afterwards explained, that this is unattended by any vibration of the larynx. By opening the mouth widely, and making a slight noise, without the expiratory movement, the sound 'ah' for the letter *a* is evoked; this being attended by a vibratory movement of the larynx which can be felt to be communicated to the fingers pressed upon it. At first the loud inharmonious animal noises that are made in attempt at speech require to be modulated. This is effected in two or three ways. The teacher himself speaking in a low tone, calls the attention of the child to the quiet, subdued motions of his chest and of the muscles around his mouth. He tightly holds the hand of the child in his own, and by depressing it, the child learns to connect this movement with a lowering of its own voice. By placing the hand of the child on his (the teacher's) throat, and by placing his (the teacher's) hand on the child's throat, he will draw its attention to the slighter vibration of the larynx when the voice is lowered. By enforced attention of this kind, the child, as his education advances, soon learns that his progress depends on his actively cultivating his powers of imitation, and by copying these movements produces in this way a fall in the voice.

"Suppose the child to have produced the sound for *a*. By filling out the cheeks and making a puff, the sound which corresponds to the letter *p* is elicited. Let these two last movements be carried on consecutively and the word *ape* is produced. The attention of the child is then at once called to the object, a picture of,

or better still, a stuffed ape. From that time forward
he connects in his mind the idea of an ape with the
sound which he has learned to make. Again, after
making the sound for *a,* he is shown the letter written
down; he then learns to write it himself, and is thus
able, first, to recognise the word when spoken by his
teacher; secondly, to speak it himself; thirdly, to
understand its meaning; fourthly, to recognise it when
written; and fifthly, to write it himself. Each of these
branches of instruction, therefore, go together hand in
hand. To perfect some of the sounds, it is necessary
to make use sometimes of certain aids; for instance, in
making the sound for the letter *p,* some children will
not compress the lips sufficiently firmly, and thus they
will produce the sound 'pooh' instead of 'pah.' By
making the child blow away a little piece of paper for
a few times, when he is making his attempt, this mistake is avoided.

"When the lips are compressed with a slow expiration and advanced, the sound corresponding to *w* in
English and *ou* in French is made. The lips closed
in the position for *w* are opened quickly with a puff
as said before, the sound for letter *p,* or opened
slowly when the sound 'bah' or that for letter *b*
will follow. We have then the three consonants *w,*
p, and *b* of our alphabet, in which the two lips are
alone employed. These for the present let us call bi-
labials.

"The upper incisors applied firmly to the lower lip,
and quickly separated, the sound corresponding to
letter *f,* or, slowly separated, the sound for *v* will be

produced.   These two letters, therefore, will, from their formation, be inciso-labials.

" The tongue being placed in apposition to the lower incisors, and the teeth closed, with a quick expiration, the hissing sound for the letters *s* follows.   The same, with a slow expiration, will produce the sound for *z*, with the difference, that in the latter case, the child's attention is drawn to the vibratory movements of the larynx, which are absent for the letter *s*.   The tongue being placed between the teeth, with a quick expiration, the letter *th* like the Greek *θ*, as in the word *thin*, and with a slower expiration as in the word *thine*, is effected.

" The tongue being placed in apposition to the upper incisors, and moved quickly away, the sound for *t* is pronounced with a quick expiration, and for *d* with a slow expiration.   These sounds or letters let us call inciso-lingual.

" The gutturals, such as *qu* or *k*, are sounded by expiration when the tongue is curved backwards and downwards, and this is assisted by a little pressure exercised on the child's larynx, between the two fingers of the teacher; or let the tongue be pressed back to the lower part of the mouth, and let the child be made to say the sound which he had learned before for *t*.   The sound for *g* is the same as for k, with the difference that there is a greater and deeper movement of the larynx.   The sound for *m* is made with closed lips, and expiring through the nose, while *n* is effected as follows :—The teeth being firmly closed, the air is prevented from escaping between them by placing the

tongue up against the upper incisors, and breathing
through the nose. For the sound corresponding to *l*,
the tongue is made to move up and down quickly
against the hard palate, the teeth being separated about
a finger's breadth. In the sound *r* the tip of the tongue
is kept vibrating against the hard palate, and the
vibrations of the larynx are felt to be distinct.

"To go back to the letter *a*, which, sounded *ah* as
in '*far*' is the first vowel taught. The mouth is
opened, the tongue kept flat, and an expiration is
made. The open mouth is then made a little smaller
by bringing the corners nearer to each other, producing
the sound *oo* as in hoop, and between these two comes
the vowel *o*. These are the first three sounds taught
in the vowels. In the sound for *e* the under jaw is
pressed forward, the larynx raised a little, and *a* is
attempted to be sounded, with the effect of producing
e, or written *ea* in the child's phonetic alphabet.

i is composed of and produced by *a* and *e* rapidly run into one.
ou   „   „   *a* „ *oo*   „   „
oi   „   „   *o* „ *e*   „   „

Lastly, the sound for *y* cannot be learned until the
sounds for vowels are perfectly mastered, but it will
then be found to be composed of the two sounds
equivalent to *e* and *a* rapidly succeeding each other
without a pause, or rather run into one. To show
this, the mute, if made to pronounce with rapidity a
sound written down as *eas* would articulate the word
'*yes*' as used by us.

"The spelling-book of the mute, therefore, will differ

in some degree from that of an ordinary child, as he will not make use of or connect in his mind the same sounds that we are accustomed to do in pronouncing the individual letters of our alphabet, but he will produce the very same word when the letters are joined together.

" To go back to the old example : *a p e* will spell *ape* for us and him alike, but each letter individually he will call by a different sound to that which we make when naming it. Strictly speaking, the mute's alphabet is more correct and less arbitrary than our own.

" The first year of the child's education is spent in reading the sounds of his alphabet and words of one syllable from the lips of the teacher and from the book, in articulating them, in writing them, and connecting the sounds he learns to produce with objects corresponding to them. It will be observed with infants and children who have the power of hearing, that the first sounds which they make are labial ones, such as sounds beginning with *m* and *p*, &c.; and these will also be found in mutes to be those most easily learned at the commencement of their education. As age advances in the one and instruction with the other, the sounds next easily learned are those made with the lips and teeth.

" It is necessary in teaching the deaf and dumb to be particular as to the way in which the expirations are made and concluded, for unless this be completely effected, the sounds will die away before they are perfectly emitted from the mouth. Indeed, one would

suppose it quite likely to happen that the child might
sometimes imitate exactly the motions of the lips,
tongue, &c., necessary for articulation, without emit-
ting any sound at all.   This, however, practically does
not occur; and more than this, the children who have
had as much as two years of this education will detect
the teacher or any one addressing them, if they should
form the words with the lips and omit the sound.
The second year's education is much the same as the
first, progressing, however, from words of one syllable
to words of two or three, and words where two or
three consonants come before or after a vowel, as in
the words *arms* or *straw*.   Here the difficulty met with
is that the child requires practice in repeating such
words quickly, the tendency being to pronounce
the consonants too separately, thus producing a sound
like *ser ter aw* instead of *straw, per lace* instead of
place.

" In words of two or more syllables, when the stress
is to be laid on one syllable, this is accomplished by
moving the child's hand quickly down as that syllable
is being pronounced.   In the second year the early
parts of arithmetic are commenced to be taught, and
it is obvious that as the education advances year by
year, no matter what the subject-matter that is being
learned, the practice of speaking and reading from
the lips or books is at the same time continued equally
for all.

" It may occur to some that the two systems of
Dactylology and lip-reading might be advantageously
combined, that the benefits of each might be received,

and their respective disadvantages left out. Let me say, once for all, that this is found to be impossible; more than this, it is of the greatest importance to check any disposition that the children may themselves evince to combine manual signs with vocal sounds. If this is once permitted, the child soon loses the power of keeping its attention undividedly fixed on the lips of the person speaking, and from that time begins to disprove in its own diction.

"The advocates of the system of Dactylology will be found, when speaking on this subject, to declare that only a certain number of children possess sufficient intelligence to learn articulation, and as instancing this, they will mention adults who are able to converse with facility on the fingers, and at the same time read from the lips and use articulate language. In all the cases of this kind which I have met with, on rigid examination I have found that they have first learned lip-reading and articulation, and that, having become proficient in this respect, they have subsequently learned to speak on the fingers. This, of course, is quite a different case to those I am speaking of. Here the finger-talking must be looked upon in the light of an accomplishment—of a second language, or as if acquired by hearing persons. The true physiological explanation of the indubitable fact, that, if finger-talking and manual signs are attempted to be combined with articulate language before the children have learned the latter method thoroughly, they will not be able to learn articulate speech, will, I think, be found to be the following.

If the child can make its wants known by making signs, although its attention may be called to the lips of the teacher, it will not be able to maintain its attention with sufficient perseverance and care to permit of the full development of the fine muscular sense inherent in the tongue, lips, palate, and throat, which full development is, I believe, essential in order to give the child the power of imitating with exactness and precision the movements which it is being taught.

So far we have been considering the case of children who are totally deaf, either congenitally, or having become so before they have learned to talk. In passing, I may mention two or three somewhat different conditions. Firstly, where children have become incurably deaf before they have learned to talk; not so deaf but that they can hear moderately loud sounds, but sufficiently so to prevent their hearing enough of what is said to allow them to acquire speech; with these the little hearing which remains will prevent the harshness and want of euphony which characterises the voice of the totally deaf mute who has been taught on this plan. Secondly, a little further degree of hearing will be of immense advantage; for besides speaking euphoniously, if a word shouted close to the ear can be understood, this will serve to correct the articulation, and the child will, of course, not take so long to learn. Indeed, the more hearing that is left, although not enough for a child to learn to speak in the ordinary way, the more quickly and better will language be acquired. Thirdly, in case of a child who has learned how to speak, and at an early age, say

seven, has lost its hearing completely by scarlet fever
or the like, it is well known that from the fact of its
not hearing others speak it will not be tempted to make
use of speech itself, will gradually depend more and
more on signs, and in a short time will have lost all
power of speaking. For such patients this system is
invaluable, for if they are taken in hand as soon as
they become deaf, they can with very little difficulty
be made to learn lip-reading, and will thus retain their
speech, or where it is beginning to be forgotten will
rapidly relearn.

The early part of the education is in the highest
degree laborious to the teacher, and, besides the perseverance required, it is actually physically hard work.
To perfect the sounds made by even quite the beginners, it is important as soon as possible to teach them
such sounds as represent words corresponding to some
visible object, or constantly recurring sensation. The
idea of the object or sensation having once become
connected in the mind of the child with the motion of
the lips by which he has learned to indicate it, he re-
peats it with great frequency and pleasure, thus giving
opportunities to the teacher to correct any faults he
may possess in pronunciation. Moreover, it constantly
happens, and I have myself observed it when paying
visits to the children undergoing this course of instruction under Mr Van Praagh, that one child who is
more advanced will correct one less so when the latter
makes mistakes. And again, from the very careful
and *prononcé* way in which the children enact the movements of the lips and tongue, the smallest variations

in these particulars are at once noticed by their con-
panions and immediately imitated. It results, then,
that a correction from the teacher which serves for the
one pupil will indirectly affect the others in the class.
Thus it is more advantageous than otherwise that
numbers should be taught under the same roof. It
must not, however, be understood from this that the
separation of the mutes from speaking individuals is
recommended. In fact, after the first year or two of
instruction, the reverse is far the better plan; for the
more the child can mix with the outside world, with
its relatives and friends, the more do its observant
powers become cultivated; and this end is assisted by
change of persons and surrounding objects which
interest it, rather than by keeping it entirely resident
with its fellows. For this reason the institution of
day schools seems to me to be likely to prove more
beneficial, besides being less expensive, than asylums
for this method of educating the deaf and dumb; the
end and aim of this system being that the children,
when grown up, should occupy positions of usefulness
where they will receive the ideas of others and communicate their own to ordinary men and women,
rather than that they, like those who only talk on their
fingers, should have occupations where their social
intercourse is limited to those similarly affected as
themselves.

Among the advantages which this method of teaching
appears to possess over that of speaking by the use of
the fingers and artificial signs, one of the chief is the
following :—That when the children have acquired the

power of talking by the dumb alphabet, however perfectly, and go out into the world, they are still deprived
of all intercourse with their fellow-creatures, excepting
in those very rare instances where they happen to meet
with those who are able to converse in the same way
as themselves, the proportion of such persons in ordinary life being so extremely small that for the sake
of argument they might almost be put out of the question; or again, on the supposition that a mute could
only acquire the power of *reading* from the lips of
others without being able himself to articulate, and
thus convey his ideas to others, it becomes a question
whether he would not, although he had to reply to
everything by writing, be a more useful member of a
community composed of ordinary speaking individuals,
than a mute who could only receive and convey his
ideas to those similarly situated as himself, or be
dependent for conversation on a chance meeting with
some one who had acquired his peculiar language.   If
such a proposition bears a moments reflection, it must
be apparent how very considerable must be the advantage of that mute who possesses not only the power of
receiving information conveyed in language familiar
to all, but also of replying in the same manner.   Both
of these faculties, however, are, in the system under
consideration, so intimately combined, that the one
naturally follows the other.

   " In the early part of the education this method,
undoubtedly, is more slow than the other in the facility
which it gives for conversing, either by the child or to
the child; but this, I submit, ought to possess very

little weight when compared with the results attained
at the end of the course of instruction.

"The condition that the child is placed in with regard
to its importance as a member of society when it leaves
school, and the careers of usefulness that are then open
to it, should, to my mind, be the chief points which
ought to be kept in view when comparing the respective merits of the two systems. Of what importance is
it if, at ten years of age, a child can talk well with its
fellows—well with those who have learned its language
—if, at fifteen years, it still only possesses the power
of communication with the few, while its companion
(brother or sister, it may be) of the same age who has
been taught on the other plan, converses by means of
and understands the language of the many? Looked
at in this light, the position of the two children hardly
admit of comparison. With regard to the ultimate
chances of success with any child of ordinary intellectual powers, M. Saegert, the Inspector-General of the
Education of Deaf Mutes in Prussia, in his report of
1856, says: 'Ninety-nine per cent. of deaf mutes
have the organs of speech normal; they will speak if
they have good sight and touch: the greater or less
probability of success depends solely (*uniquement*) on
the greater or less capacity of the master.'"

From what I have said you will perceive at once the
estimation in which I hold either system of education
for dumb children. And it remains now for you by
personal observation to satisfy yourselves of the fal-
laciousness or correctness of my views of this question,

I have only to add that when a child of five, six,

seven years of age or upwards loses its hearing, if it is
able to read, it should be made to do so without any
delay two or three times during each day, for by such
an exercise it will retain all words that it has previously
known, and so will be prevented losing speech. Its
vocabulary will have to be increased, and the rest of
its education conducted by instruction upon the method
which I have referred to.

The conclusion of the subject of deaf-mutism brings
to an end this course of lectures.

INDEX

PRINTED BY J. E. ADLARD, BARTHOLOMEW CLOSE.

16